SAMPLING
THE GREEN WORL

SAMPLING
THE GREEN WORLD

Innovative Concepts of Collection,
Preservation, and Storage
of Plant Diversity

EDITED BY
TOD F. STUESSY AND S. H. SOHMER

COLUMBIA UNIVERSITY PRESS
NEW YORK

Columbia University Press
New York   Chichester, West Sussex
Copyright © 1996 Columbia University Press
All rights reserved

Library of Congress Cataloging-in-Publication Data
Sampling the green world : innovative concepts of collection,
    preservation, and storage of plant diversity / edited by Tod F.
    Stuessy and S.H. Sohmer.
        p.   cm.
    Based on a symposium, sponsored by the Botanical Research
    Institute of Texas and the Herbarium of the Ohio State University,
    held in Fort Worth, Tex., Nov. 19–20, 1993.
    Includes bibliographical references and indexes.
    ISBN 0–231–10136–8 (cl.)
    1. Plants—Collection and preservation—Congresses.   I.Stuessy,
    Tod F.   II. Sohmer, S.H.   III. Botanical Research Institute of
    Texas.   IV. Ohio State University.   Herbarium.
    QK61.S26   1996   96-3600
    579—dc20

Casebound editions of Columbia University Press books are
printed on permanent and durable acid-free paper.
Printed in the United States of America
c 10 9 8 7 6 5 4 3 2 1

CONTENTS

PREFACE

This book originated from a symposium of the same name, sponsored by the Botanical Research Institute of Texas (BRIT) and the Herbarium of The Ohio State University, held in Fort Worth, Texas, 19–20 November 1993. The origin of the symposium developed from a perceived need for addressing issues relevant to methods of collecting, preserving, and storing plants. The systematic botanical community has already put considerable emphasis on defining priorities for inventorying plant diversity in different parts of the world. Even within tropical regions attempts have been made to determine which countries need the greatest attention (e.g., Campbell and Hammond 1989). Likewise, the new Systematics Agenda 2000 Report (Anonymous 1994) speaks loudly and clearly for the need to inventory plants (and other organisms) in the tropics as well as in other poorly known habitats (e.g., the deep ocean).

Botanical and other systematists have also put considerable attention on dealing with data from our existing collections, including formats, capabilities for sharing such information, and ethics of access (Brenan, Ross, and Williams 1975; Allkin and Bisby 1985; Duncan 1991). Many valuable perspectives on these topics also emanated from the Flora of North America Workshop held in 1988 in Alexandria, Virginia (Morin et al. 1989). With computers being more accessible and increasingly a part of our daily lives, these issues will be even more significant in the years ahead.

The systematic botanical community, therefore, has already given attention to where to collect, and what to do with data already existing on specimens in our collections. This book does not deal with either of these important issues. Rather, it focuses on the middle area that so far has not been considered in detail, that is, the collecting of botanical materials, their preservation, and their storage for the future. The Society for the Preservation of Natural History Collections (SPNHC) has begun to address some of these issues, such as were discussed in the symposium in Madrid in 1992. Some publications have also appeared that bear on these topics. The useful volume *Storage of Natural History Collections: Ideas and Practical Solutions* (Rose and de Torres 1992) contains many helpful ideas, although covering a broad range of curatorial areas (anthropology, zoology, and so on). The important volume *Conservation of Plant Genes* (Adams and Adams 1992) is most appropriate for addressing issues of DNA storage. This is a significant topic (covered by two papers in this book), as herbaria can now be correctly viewed as DNA repositories as well as historical archives. Our own perception of herbaria is changing, and we need to adjust our management and development activities accordingly.

It would be useful to know if the materials we are accumulating now in the world's herbaria are the right kinds of specimens, in sufficient quantities, with the proper preservation techniques, and with the proper storage environments so that they will serve societal needs into the next century and beyond. With the disappearance of so many native species of plants, and with the accelerating rate of this extinction into decades ahead because of increasing human population pressure, it is imperative that we address these issues now, while there is still a chance to do so. Whatever we already have in our collections, plus whatever we may be able to acquire in the next fifty years, may well be the last information about many of these life forms. Are we gathering the right information about them and are we storing it for maximum future utility?

This book begins with a historical overview of documenting phytodiversity (Vernon Heywood), and societal and scientific information needs from plant collections (Brian Boom), that is, what have we been doing up to this point, and what are the societal expectations for us and herbaria that might guide us into the future? To put this another way, basically why does society believe we are needed and how can we best serve this need?

The book then addresses the issue of sampling of plant diversity, beginning with statistical adequacy of collections we now have or intend to acquire in the future (Bernard Baum). Basically, do we have what we want and need in order to answer the significant questions involving information content of classifications, phylogenetic reconstructions, and insights regarding the evolutionary process? James Miller outlines protocols for collecting samples for pharmaceutical research (i.e., secondary plant products), so important in the development of new drugs.

Sometimes it is physically impossible to sample particular plant materials, but it may be feasible to obtain images of them. The basic question is this: realizing that in some cases it will be impossible, inconvenient, or undesirable to procure the actual plant materials, what are the options for storing images of this diversity, and to what extent should herbaria be devoted to these efforts? Obviously we have used botanical art and illustrations for decades (as we read in the chapter by James White), then photography, and now computerized video applications involving various technologies even reaching into virtual reality (chapters by David Kramer and Don Stredney, respectively). All this is being complemented by Geographic Information System (GIS) techniques, which can enhance understanding of environmental correlations with plant taxa (see chapter by Michael DeMers). To what extent can we, or should we, preserve images of these data and how might herbaria respond to these challenges? An ancillary, difficult question is to what extent specimens themselves are really no longer useful, or at least not as meaningful to as many people in as many places as they used to be?

Storage of collected plant materials is also a significant activity in systematic botany. This book addresses pollen, wood, and other pickled materials (Sherwin Carlquist), DNA from different perspectives (Robert Adams; Dennis Loockerman and Robert Jansen), and techniques for extending the life of any type of plant tissue (Eric Roos et al.). We need to ask if there is, or might eventually be, a universal solvent or technique that lets us get all the information in one type of preservation or storage? Should herbaria become more directly involved in DNA (or other) storage efforts? Should every herbarium attempt to preserve DNA in the same way or should we develop a cooperative strategy with some institutions specializing in one set of techniques and others with alternative methods? If so, how do we coordinate this?

The final issue covered in this book is how to deal with already-stored materials. Duplicate herbarium specimens have been a perennial topic of discussion, especially in the offices of funding agencies. William Anderson addresses this topic directly to put on record a community perspective. Do we wish to classify (or grade) the types of storage environments in which we are storing these materials (chapter by Christine Niezgoda)? Such suggestions may distress directors of collections, fearful of funding agencies using these criteria as bases for future collections support. On the other hand, this might be a useful way to encourage institutions to upgrade existing environments.

The issues addressed in this book are of great importance to ourselves and future generations. If we can anticipate societal needs and respond to them with an eye to the future, we will maximize our contributions as plant biologists dedicated to studying systematic and evolutionary relationships within the Green World.

T. F. STUESSY

S. H. SOHMER

Literature Cited

Adams, R.P. and J.E. Adams, eds. 1992. *Conservation of Plant Genes: DNA Banking and In Vitro Biotechnology*. San Diego: Academic Press.

Allkin, R. and F.A. Bisby, eds. 1985. *Databases in Systematics*. London: Academic Press.

Anonymous. 1994. *Systematics Agenda 2000: Charting the Biosphere. Technical Report*. New York: Systematics Agenda 2000.

Brenan, J.P.M., R. Ross, and J.T. Williams, eds. 1975. *Computers in Botanical Collections*. London: Plenum Press.

Campbell, D.G. and H.D. Hammond, eds. 1989. *Floristic Inventory of Tropical Countries: The Status of Plant Systematics, Collections, and Vegetation, Plus Recommendations for the Future*. New York: The New York Botanical Garden.

Duncan, T. 1991. *Development of a Specimen Management System for California Herbaria (SMASCH)*. Assoc. Calif. Herbaria Working Paper no. 1: Berkeley, California.

Morin, N.R., R.D. Whetstone, D. Wilken, and K.L. Tomlinson, eds. 1989. *Floristics for the 21st Century*. St. Louis: Missouri Botanical Garden.

Rose, C.L. and A.R. de Torres, eds. 1992. *Storage of Natural History Collections: Ideas and Practical Solutions*. Pittsburgh, Pa.: Soc. Preserv. Nat. Hist. Coll.

ACKNOWLEDGMENTS

This book could not have been produced without the help of many individuals. Foremost have been the contributors who accepted the challenge of pulling together ideas and presenting them, both in the Fort Worth symposium and again here in book form. They have offered new ideas and challenged the plant systematics collections community to be forward-looking. Without these creative insights, as well as attention to practical matters such as symposium and editorial schedules, this book could never have been compiled.

We are most grateful for funding from the Sid Richardson and Bass Foundations, which made the symposium possible and led to the completed book project. Support materialized at a very crucial stage and was key to keeping the symposium on schedule.

During the symposium, the staff of the Botanical Research Institute of Texas (BRIT) gave special service to ensure that all went successfully. We wish to thank Susan Richardson for her extraordinary efforts in dealing with accommodations, meeting rooms, audiovisual equipment, and airline schedules, made even more exasperating by a strike at American Airlines, which required rescheduling of tickets for several speakers and delays for others. Without Susan's help, the symposium and this book would never have resulted. Barney Lipscomb also gave substantial service to ensure that the symposium went as planned.

We also thank the editorial and production staffs of Columbia University Press, in particular Alissa Bader, Anne McCoy, and Kerri Cox for their outstanding help through many difficulties of book preparation. Rita Bernhard provided excellent copyediting. We also appreciate the confidence in the project radiated by Ed Lugenbeel, who helped shepherd it from rough idea to polished bound volume.

Finally, we thank Larissa Menon and Ulf Swenson for careful reading of the proofs, and Linda Thornton for preparing the index.

CONTRIBUTORS

ROBERT P. ADAMS
Biotechnology Center
Baylor University
Waco, TX 76798

WILLIAM R. ANDERSON
Herbarium
University of Michigan
Ann Arbor, MI 48109

BERNARD R. BAUM
CTR Land & Biological
Resources Research
Department of Agriculture
Central Experimental Farm
Ottawa, ON K1A 0C6
Canada

SHEILA A. BLACKMAN
National Seed Storage Laboratory
U.S. Department of Agriculture
Agricultural Research Service
Ft. Collins, CO 80521

BRIAN M. BOOM
New York Botanical Garden
Bronx Park, NY 10458

SHERWIN CARLQUIST
4539 Via Huerto
Santa Barbara, CA 93110

MICHAEL N. DEMERS
Department of Geography
New Mexico State University
Las Cruces, NM 88003

VERNON H. HEYWOOD
Department of Botany
University of Reading
Whiteknights, RG6 2AS
Great Britain

ROBERT K. JANSEN
Department of Botany
University of Texas
Austin, TX 78713

DAVID W. KRAMER
Department of Plant Biology
The Ohio State University
Mansfield, OH 44906

DENNIS J. LOOCKERMAN
Department of Botany
University of Texas
Austin, TX 78713

JAMES S. MILLER
Missouri Botanical Garden
P.O. Box 299
St. Louis, MO 63166

CHRISTINE J. NIEZGODA
Department of Botany
Field Museum of Natural History
Chicago, IL 60605

ERIC E. ROOS
National Seed Storage Laboratory
Agricultural Research Service
U.S. Department of Agriculture
Ft. Collins, CO 80521

S. H. SOHMER
Botanical Research Institute of Texas
509 Pecan Street
Ft. Worth, TX 76102

PHILIP C. STANWOOD
National Seed Storage Laboratory
Agricultural Research Service
U.S. Department of Agriculture
Ft. Collins, CO 80521

DON STREDNEY
Ohio Supercomputer Center
The Ohio State University
Columbus, OH 43212

TOD F. STUESSY
Natural History Museum of Los
Angeles County
900 Exposition Boulevard
Los Angeles, CA 90007

LEIGH E. TOWILL
National Seed Storage Laboratory
Agricultural Research Service
U.S. Department of Agriculture
Ft. Collins, CO 80521

CHRISTINA T. WALTERS
National Seed Storage Laboratory
Agricultural Research Service
U.S. Department of Agriculture
Ft. Collins, CO 80521

JAMES J. WHITE
Hunt Institute for Botanical
Documentation
Carnegie-Mellon University
Pittsburgh, PA 15313

SAMPLING
THE GREEN WORLD

PART 1

The Importance of
Documenting Phytodiversity

How can we hope to manage and preserve plant diversity on this planet if we have no idea of the resources that actually exist? As Vernon Heywood states so clearly in chapter 1, the plant taxonomic community has attempted for more than two hundred years to organize information about plants on Earth. But despite these laudatory efforts, we still are uncertain on how many species of plants exist. There are approximately 250,000 flowering plants and ferns known, but numerous species still await description and more in-depth investigation. Stuessy (chapter 16) estimates that at least 50,000 more plant species remain to be discovered. Heywood suggests that we may have catalogued only 15% of the total plant and animal diversity so far. In any event there is still much left to do.

Many lessons can be learned from the past. As Heywood (chapter 1) shows us, over the years there has been a normal preoccupation with chronicling diversity, but without focusing on details of documentation. This has been particularly notable in botanical gardens, where many accessions have been inadequately labeled as to place of origin, and hence have been of limited research value. Herbarium material is mixed with regard to precision of information, with specimens even from this century sometimes lacking details of location, ecology, and plant features. The complex nature of our plant taxonomic literature has also required scholasticism to deal with past observations published in obscure places. To document phytodiversity effectively, we must do a better job with data we judge important.

Brian Boom in chapter 2 shows convincingly how dependent we are on botanical materials. There is almost no aspect of human activity—from foods we eat, to gasoline (fossil plant products), to plastics, beverages, and clothing—that doesn't directly or indirectly involve plants. As systematic botanists, we have been particularly negligent in getting this message to the ordinary citizen, who supports our activities. We frequently complain that we are underfunded, but few of us spend time talking to legislators, influential citizens of our community, and interested garden groups. People do not respond to existing needs; they respond to their own beliefs that needs exist. Someone has to provide them with information to internalize these beliefs—and that someone must be us. We have to reach out better to the general public.

The importance and urgency of documenting phytodiversity, the first step in conserving valuable plant resources, is described clearly in chapter 3 by S. H. Sohmer in the context of the Philippine Islands Flora Project. More than 85% of Philippine forests have been cut for various needs, and present destruction of forest continues at an alarming rate. Likewise, no sustained plant taxonomic work force has existed in the country, resulting in a real lack of understanding of plant resources. Highlighting rare and endemic plants is essential to allow proper management and sustained use of plant resources. The new Philippine Islands Flora is a good model of modern cooperative projects, which focus not only on plants but also on training of new plant scientists to carry on study and management after the project terminates. Further, the Philippine Islands Flora Project stresses databasing of information to serve the interests of Philippine society long after the hard copy flora appears. We need more projects of this nature.

1

A Historical Overview of Documenting Plant Diversity: Are There Lessons for the Future?

VERNON H. HEYWOOD

Attention is drawn to the fact that during much of the history of collecting and storing material and information about plants ("sampling the green world"), most emphasis has been given to (1) living plants grown in botanic gardens; (2) material preserved in herbaria and museums; and (3) books, papers, and illustrations in libraries and archives. It is noted that today our perception and means of sampling the living world are much more sophisticated, as are our ways of documenting and storing the data we gather and communicating about them. A historical review is given of the ways in which plants were sampled in the eighteenth and nineteenth centuries, so as to help us understand current practices, many of which are enshrined in past traditions and perhaps shed light on some of the problems we are facing today. It is pointed out that so far we have only managed to name about 15% of the species of organisms believed to exist on our planet, so the task still before us is colossal. It is stressed that the environmental crisis facing the world today is so serious that we must urgently seek the most effective means at our disposal, including the latest technologies, to sample, document, save, study, and use wisely as much plant (and animal) diversity as possible—in accordance with the message of the Convention on Biological Diversity signed at Rio de Janeiro in 1992, which came into effect at the end of 1993.

For much of its history the collection and storage of the green world, to use the title of this book, has been dominated by three aspects:

- living plants brought into cultivation in botanic gardens;
- preserved material in herbaria and museums; and
- books, papers, and illustrations in libraries and archives.

Today in the post-UNCED (United Nations Commission on Environment and Development) world our perception of plant diversity is much wider and ranges from alleles and genes through populations to species and higher taxa in ecosystems and landscapes. We are also concerned with social and economic aspects of plant diversity and engage in the sampling, storage, and documentation of plant genetic resources and in sampling for ethnobiological studies. The kinds of materials we collect and store for taxonomic and systematic studies are much wider than before and often involve highly sophisticated techniques as, for example, in the assessment of genetic variability by RFLPs, RAPDs, and microsatellites. Also with regard to the documentation, handling, communication, and storage of data, we are in the midst of a technical revolution and where it will end we can only hazard a guess. It would be a brave scientist who would predict where we will be even ten years from now in terms of virtual reality in this postmodernist world! (See the conclusion of this volume for a vision of what this future might be.) Of course there is little in the past history of botany to inform and advise us in terms of these modern technical developments, although I believe there are lessons for taxonomy and systematic biology in the areas of plant collections, both living and preserved, literature and our perception of the natural world.

The Dimensions of Plant Diversity

One of the most significant developments in plant systematics was the recognition of the vast numbers of taxa that exist in nature. This was a gradual realization and even today we are still not sure how many species occur in most groups of plants, as will be discussed later. If we consider the beginnings of scientific taxonomy in the mid-eighteenth century, horizons were limited, concepts of variation extremely limited, and the numbers of botanists engaged in what today we call taxonomic studies very small. The first global account of plant species was Linnaeus's *Species Plantarum* (1753),which contained about 1,350 genera and 8,550 species. In essence it was a codification of European existing folk taxonomies, and as I have suggested on a previous occasion (Heywood 1980), it was only possible for Linnaeus to accomplish this overall synthesis down to the species level because of the relatively small number of species that had been described, albeit in the chaotic literature of the time. Later, the increasing flow of new species and genera being described would perhaps have deterred Linnaeus from even starting the task.

The state of exploration of plant and animal resources of the world at the time of Linnaeus and his contemporaries was so superficial as to allow him to undertake his codification. Parts of Europe, especially the west and south, were quite well known floristically, but plants from other continents and the tropics were poorly represented in herbaria or in botanic gardens, although some parts of the world, including North America and India, were being opened up by explorers and colonizers. The great phase of imperialist expansion and discovery of the nineteenth century, leading to the creation of the great European cathedrals of taxonomy—such as the herbaria and museums of Berlin, Leiden, London, Paris, Prague, Stockholm, and to a lesser extent those in the New World—was yet to come.

Another, more detailed assessment is given by Adanson in the introduction to his seminal work *Familles des Plantes* (1763). Adanson had a much greater appreciation of the state of knowledge of plant life in the world. Unlike Linnaeus, who had not ventured beyond Western Europe and Lapland, Adanson had visited the tropics in Senegal and observed shrewdly that "really, botany seems to change face entirely as soon as one leaves our temperate countries in order to enter the torrid zone" (Stafleu 1963:179). Although Linnaeus, in the introduction to *Species Plantarum* (1753), considered that the number of plant species in the world would scarcely reach 10,000, Adanson calculates that 18,000 species were then known and 25–30,000 still to be described. He also made the observation that of the 18,000 species known, only 3–4,000 were tolerably well known—a remarkable insight in those days and reflecting his philosophical approach, which was totally opposed to that of Linnaeus. Adanson also notes that from classical times until 1763, there are 2,000 authors of botanical publications and 4,000 volumes of which only half deal with botany sensu stricto. The number of illustrations of plants published before 1759 is estimated by him at 70,000 illustrating 10,000 species but only 1,500–2,000 satisfactorily.

Even today, we are dramatically ignorant about even such basic facts as the numbers of species in particular countries or regions of the world. There is general agreement that the overall numbers of flowering plants and ferns is on the order of 250,000. Precisely how this figure has been derived is not clear, but I have been able to confirm it by totaling the numbers of species given for the different families in *Flowering Plants of the World* (Heywood 1978, 1993a), which gave some 230,000-plus species. Of course the family totals are little more than estimates in most cases. It was surprising, therefore, in reviewing the statistics held by the World Conservation Monitoring Centre and the data given in the Regional Reviews prepared as part of the IUCN/WWF project *Centres of Plant Diversity* (Davis and Heywood 1994) to find that the total numbers of single-region endemics reached 235,703. This would suggest that the total number of all species of flowering plants and ferns is well over 250,000 once one takes into account those species that have wider distributions. I have no solution to offer at this stage but simply wish to underline just how poorly based some of our basic elements of plant documentation are. It should be noted, too, that for some countries only approximate numbers of species can be given, such as Bolivia, Brazil, and Papua New Guinea. Likewise, the numbers of endemic species in some countries can only be guessed.

Totals for the tropical floras of the different continents are also subject to major revisions. Thus the flora of tropical Africa was estimated until recently at 35,000 species but is now considered to be no more than 21,000 (Stork, in WCMC 1992), about the same as the flora of southern Africa.

I have focused on this period in the mid-eighteenth century since it is when ordered scientific taxonomy began, and much of what we do in taxonomy today is still influenced by attitudes established during that time. We are prisoners of our past in many regards (Heywood 1983). On the other hand, little remains of the structure and methodology of Linnaean taxonomy, apart from the binomial system and much of the terminology, and of course the considerable number of species actually recognized or described by Linnaeus. Most of the Linnaean philosophy is rejected today (as it was by Adanson) such as the concept of special creation, the fixity and essentially invariant nature of species, and the a priori methodology. While Linnaeus's ideas, philosophies, and procedures were largely instrumental in laying the foundations and indeed made possible the development of scientific taxonomy, Linnaeus marked the end of an epoch (Heywood 1985).

Yet the Linnaean system of codification has instilled in us the desire to fit all the variation patterns we observe in nature into the straitjacket of species and genera, even in groups where evolution has not yet led to clearly distinct and definable lineages or clades. It has forced us to use a nomenclatural system and starting point from a period of unsatisfactory literature and herbarium representation, thereby encouraging museum scholasticism (Heywood 1974). It has also established the image of taxonomy as one of scholars poring over old books and dried herbarium specimens, a perception that subsequent developments did little to alter.

The Rise of the Herbarium Taxonomist

The post-Linnaean expansion of taxonomy was marked by remarkable industry and achievement in cataloguing the vast amount of new material that flowed into herbaria and museums from all corners of the globe. In the absence of any guiding genius, there seemed to be little overall plan or purpose. The expansion of the herbarium was an essential part of the new empirical taxonomy that was being developed—the study of unknown plants and animals rather than the systematization and codification of knowledge of organisms already known. It also led to the creation of the professional taxonomist, today often called the herbarium taxonomist, who was engaged primarily to describe, name, and publish books and papers about the results of the study of herbarium specimens. Increasingly these herbarium taxonomists found themselves working on herbarium specimens collected by others without ever having any direct experience in the field of the plants or animals concerned. This herbarium approach, when linked with the description of plants in isolation from their ecology, has been described by Jacobs (1980) as divorcing plant taxonomy from the living world of nature and from the applied biological sciences.

The effect of the herbarium sheets on the standardization of specimens was often quite serious. As Federov (1977) noted it probably explained the apparent homogeneity of many widespread species over vast areas. As he says, "The uniformity of herbarium specimens, sometimes accumulated in large numbers, can be explained by the fact that many collectors involuntarily discarded specimens as inconvenient on account of their dimensions or some other characteristics and thus made collections of standard specimens suitable for drying and mounting on standard sheets of herbarium material." It is in fact quite remarkable how plant taxonomists have by and large contrived to reduce the plant world to what can be accommodated on sheets or cards about 16.5" x 11.5", even allowing for the two or three sheets that are sometimes preserved for more bulky material. Yet the system was highly successful, at least in meeting the limited goals of this kind of taxonomy. Even today, taxonomists must spend large periods of time comparing herbarium specimen with herbarium specimen, and relatively small periods making comparisons with living material. Yet seldom has one questioned just how effective herbarium material is as a means of storing information on plant diversity, and whether there is an alternative in these days of high technology. The disadvantages of using herbarium material as the basis of taxonomic comparison and description are particularly acute in tropical plants, especially trees. The description of most tropical trees has been based on fine details of their leaves, buds, flowers, and fruits, and on their anatomy to a lesser degree. On the other hand, it is only in the last forty to fifty years that field characters such as bark, habitual mode of branching, and structure have been studied. It is significant that much of our knowledge of the structure and biology of tropical plants has come from research undertaken in botanic gardens such as in Singapore and Peradeniya, as pointed out by Purseglove (1959) and Holttum (1970).

There is an urgent need to take a new look at the ways material is collected for taxonomic studies and how herbarium specimens can be regularly supplemented by other forms of sample and information. The quality of material collected and its documentation was variable even in the post-Darwinian period when the significance of variation and variability was beginning to be appreciated. It is curious that although other branches of biology adopted more or less rigorous standards for the sampling and collection of field data, as for example ecology and population genetics, taxonomists still continued their somewhat random procedures.

Even today the sampling, collection, and documentation of material collected in the field by taxonomists are seldom scientifically structured and still does not generally follow any norms such as those introduced for the surveying, sampling, collection, and documentation of germplasm by genetic resource collectors, such as passport data, and ecogeographical surveying techniques. True, many institutions or individuals today use preprinted field notebooks that outline fields of information to be recorded, but there is no uniformity of approach, either generally or in respect to particular groups. Just as the International Plant Genetic Resources Institute (IPGRI) and other genetic resource institutions determine what passport data and descriptors should be employed for particular crops, perhaps we should consider something similar for collecting and annotating material to be collected of selected plant families and genera. Present-day field methods of taxonomy restrict our knowledge to the limited material and data that are recorded for the majority of species, and we know virtually nothing of their demography, detailed distribution, population variation, genetic variability, gene flow, and so on.

Not surprisingly, there have been many calls to "complete the inventory" and, as Raven and Wilson (1992) note, there is growing recognition of the need for a crash program to map biodiversity in order to plan its conservation and practical use. They propose a fifty-year plan for biodiversity surveys as a series of successive ten-year plans. Some progress has been made by the taxonomic and biodiversity communities in this general area, such as the development of rapid inventory techniques that have been applied in various parts of the world, the proposals for All Taxa Biodiversity Inventories (ATBI) by Janzen and colleagues, and the inventorying and monitoring component of the *Diversitas* program (di Castri, Robertson Vernhes, and Younès 1992; Robertson Vernhes and Younès 1993), but in general field collecting techniques by taxonomists have not changed much over the last hundred years.

The Biosystematic Phase

The development of what came to be known as biosystematics in the 1940s led to a great emphasis on the dynamics of systematics, and treated species for the first time as a series of variable evolving populations representing different kinds of evolutionary situations. In terms of information gathering, the emphasis was on intensive sampling in the field or later in the laboratory of material or data taken from populations. Because of the time and costs involved, the numbers and geographical range of populations sampled were small, with the result that it was difficult to draw any generally applicable practical taxonomic conclusions from the material.

One of the commonest approaches during the biosystematic phase was the study of karyology, especially chromosome numbers. Here not even detailed sampling was made within populations, let alone between them, in different parts of the species' range. For only a small number of taxa were extensive surveys carried out and, in the absence of these, single counts were often all that one had to represent a species (still true for many), and this led to the perpetuation of a series of myths about the relationship between chromosome number and taxonomy.

Similar problems arose with the use of various analytical techniques in what came to be known as chemosystematics or molecular taxonomy. Even today, most molecular data are taken from very small samples, and it is not possible to assess the full range of variability in chemical features used in taxonomy, and consequently their validity.

A general lesson that can be drawn from these various illustrations is that when applying new techniques, thorough sampling must be stressed. It has taken taxonomists two centuries to appreciate fully the dynamic nature of species and their populations, and it is not surprising therefore that more experimentally inclined scientists are somewhat cavalier in their approach to variation and variability. Physiologists and biochemists are remarkably typological in their approach, often not even bothering to ascertain the origin and identity of their material, let alone appreciating the possibility that different samples might give different results.

Botanic Garden Collections

Until recently, a similar failure can be observed in adequately documenting much of the material brought back into cultivation in botanic (and some private) gardens. The origin of plants in nature was often confused with their source. Thus many botanic gardens supplied one another with seed and other propagules through the Index Seminum seed-exchange system instituted in the eighteenth century, and in countless cases the origin is given as the botanic garden from which the material was obtained and the original native source lost (Heywood 1976). What has been termed a *botanic garden flora* developed, with the same array of species, usually those that were most easily cultivated, adaptable to a range of different conditions, tolerant of a degree of neglect, and represented by material probably of the same origin, usually to the neglect of the local flora (Cullen 1976). Failure on the part of botanic gardens to appreciate the need for scientific documentation of the material grown and for growing an adequate range of samples of a particular species has led to a situation whereby much of the vast collection of material in cultivation in the world's 1,600 plus botanic gardens has little value as conservation material (Heywood 1990).

Until recently few botanic gardens had adopted an accessions policy, although this is now recognized as a key instrument in their planning, development, and maintenance (WWF/IUCN/BGCS 1989). A survey of North American botanic gardens by Vandiver (1988) showed that 53% of 26 gardens responding to a survey had written collection policies, while an earlier survey by Correll (1980) found that only 10% of 145 gardens had formal collections policies. Moreover, gardens do not always follow their collections policy when they do have one! Attempts should be made to rationalize collections, especially at a national level, so as to avoid unnecessary and expensive duplication. A lead has been given by the Dutch Botanic Gardens Foundation, which was formed to coordinate and facilitate activities of the botanic gardens already cooperating in the Decentralized National Plant Collection (de Jong 1984).

Botanic gardens are often very poorly informed of what they have in their collections, and the statistics they issue are often highly unreliable, frequently confusing numbers of species with numbers of taxa, including cultivars, as I have been able to confirm in many instances.

On the other hand, many botanic gardens have significant accessions of wild-collected material, often of considerable conservation value, although usually in very small samples of one or a few plants (Heywood 1990). It is only in recent years that the need for the use of rigorous techniques for the sampling, documentation, and maintenance of scientific collections in botanic gardens has been appreciated following the prescriptions outlined in *The Botanic Gardens Conservation Strategy* (WWF/IUCN/BGCS 1989), especially by those gardens that have adopted a more explicit role in the collection of germplasm of wild species and cultivars (Heywood 1993b).

One area where botanic gardens have made significant progress, and are in some ways ahead of herbaria, is in the development of record systems and of international formats for the exchange of records. *The International Transfer Format for Botanic Garden Plant Records* was published in 1987 by the Botanic Gardens Conservation Secretariat and is now used by botanic gardens around the world for the exchange of computerized information about their plant holdings.

What lessons can we learn from the past history of botanic gardens? Perhaps the principal one is that we need to be less indulgent in our approach to collecting and growing plant diversity and should take much more care in what we grow, why we grow it, how we sample it, how we use it, what conservation or scientific value it might have, and how we make it available for different user publics. It is a privilege to hold plant collections—not a right. We must also take into consideration intellectual property rights and other concerns and legitimate interests of the countries from which we collect material. For too long we have adopted a somewhat colonialist attitude to collecting plants, both living material and herbarium samples. Increasing recognition of the economic potential of many plant species has caused a change in attitude which is increasingly leading to national legislation safeguarding countries' interests in their own plant resources, and this is also reflected in the *Convention on Biological Diversity*, which came into effect at the end of 1993.

Cataloguing the Literature of Plant Diversity: The Paper Chase

Until recently, the prowess of a taxonomist was judged as much by his or her knowledge of the literature as by his or her achievements in publications or field knowledge; indeed some of us were regarded as walking databases! We have since moved rapidly into an electronic age whereby personal computers, modems, electronic mail, and faxes sit on the work bench, even though the realm of systematics was where the computer was slowest to invade (Heywood 1984).

Shetler (1974), in a seminal paper, suggested that this time lag in the use of computers by taxonomists, even in the face of colossal information-processing and retrieval problems, was probably a reflection of "the constitutional aversion of the scientist to slowing up the pace of primary discovery for the sake of organizing and synthesizing what is already discovered."

Already in 1977, as an appreciation was dawning of the size of the diversity problem and the scale of losses anticipated as a result of human action such as deforestation of tropical forests, Raven was writing, "If we are truly interested in cataloguing this diversity . . . it seems all but incomprehensible that we have not made use of electronic data processing to keep track of the units being classified and allow the much more efficient accumulation of information about them. There is literally no other way in which this can be done." After a slow start, a number of major taxonomic database and information systems have now been initiated, including the International Legume Database and Information System (ILDIS), the International Organization for Plant Information (IOPI), the European Taxonomic, Floristic, and Biosystematic Documentation System (ESFEDS), the new Flora of North America Project, and Plantas do Nordeste (PNE). Today, with new computer technologies such as multimedia, client-server architecture, the Pentium chip, parallel processors, image analysis, and artificial intelligence, the prospects are exciting (Allkin and Winfield 1993).

All this is a far cry from the days, not so long ago, when taxonomic data were published (often in Latin) in dense, virtually unreadable tomes that depended more on the "authority" of the authors than on the data presented. Indeed the precise data to substantiate taxonomic decisions were not usually published, although specimens consulted were sometimes cited. Again, what can we learn from the past?

I believe there are three main lessons. First, the value of data, whether in electronic form or in notebooks, depends on their accuracy. As Allkin and Winfield (1993) rightly point out, however valued the opinions of its author, a data set is worthless if one cannot rely on the accuracy of the observations—nor, one might add, if one cannot depend on the *source* of the observations (which invalidates, as noted above, many sets of chemical data). Second, we need to talk a common language so as to be able to communicate effectively. Data standards are needed, and they must be adopted. It is remarkable that the Taxonomic Databases Working Group, which I was instrumental in founding, dates from as late as 1985. It has subsequently issued a series of data standards, ranging from international exchange formats to geographical, bibliographic, and plant-use data. Third, although taxonomists have tended in the past to ignore their actual and potential consumers and have directed most of their publications at one another, this is no longer a road that can be followed. Taxonomists have to learn to communicate effectively with the users of taxonomic products and present their data in a form suitable for their needs. User-friendly publications are perfectly possible with today's electronic publishing media.

We have to get the data out to the users, wherever and whoever they may be, using VideoFloras, CD-ROM technology, or, before long, virtual reality.

Only in the past two or three decades have we come to realize that plant (and animal) diversity is so great that we have only managed to name about 15% of the species believed to exist on our planet (Wilson 1992); at the same time we have begun to appreciate that our mismanagement of much of the more conspicuous diversity—such as higher plants, birds, and mammals, and the forests, grasslands, wetlands, scrublands, and deserts in which they live—is leading to its loss on such a scale that we cannot afford to neglect its conservation. How we set about sampling and monitoring this remaining diversity will be critical in allowing us to save, use, and study it wisely.

Literature Cited

Adanson, M. 1763. *Familles des Plantes*. 2 vols. Paris.

Allkin, B. and P. Winfield. 1993. Cataloguing biodiversity: New approaches to old problems. *Biologist* 40:179–183.

Botanic Gardens Conservation Secretariat. 1987. *The International Transfer Format for Botanic Garden Plant Records*. Hunt Institute for Botanical Documentation: Pittsburgh.

de Jong, P. C. 1984. Towards a Decentralized National Plant Collection in the Netherlands: Specialization as a survival strategy for plants and gardens. In P. Maudsley, ed., *Proceedings of the First International Conference. European Mediterranean Division of the International Association of Botanical Gardens*, pp. 63–68. *Reports from the Botanical Institute, University of Aarhus*, no. 10.

di Castri, F., J. Robertson Vernhes, and T. Younès, eds. 1992. *Inventorying and Monitoring Biodiversity: A proposal for an International Network. Biology International*, special issue no 27.

Correll, P. G. 1980. *Botanical Gardens and Arboreta of North America: An Organizational Survey*. Los Angeles: American Association of Botanical Gardens and Arboreta.

Cullen, J. 1976. The use of records systems in the planning of botanic gardens collections. In J. B. Simmons, R. I. Beyer, P. E. Brandham, G. Ll. Lucas, and V. T. H. Parry, eds., *Conservation of Threatened Plants*, pp. 95–103. New York: Plenum Press.

Davis, S. D. and V. H. Heywood. 1994. *Centres of Plant Diversity: A Guide and Strategy for Their Conservation*. Vols 1–3. Gland, Switzerland: WWF and IUCN.

Federov, A. A. 1977. On speciation in the humid tropics: Some new data. *Gardens' Bulletin Singapore* 27:129–136.

Heywood, V. H. 1974. Systematics: The stone of Sisyphus. *Biol. Journal Linn. Soc.* 6:169–178.

——. 1976. The role of seed lists in botanic gardens today. In J. B. Simmons, R. I.

Beyer, P. E. Brandham, G. Ll. Lucas, and V. T. H. Parry, eds., *Conservation of Threatened Plants*, pp. 225–32. New York: Plenum Press.

——. 1978. *Flowering Plants of the World.* Oxford: Oxford University Press.

——. 1980. The impact of Linnaeus on botanical taxonomy: Past, present, and future. In Carl von Linné, *Uber Zeitgeist, Werk und Wirkungsgeschichte*, pp. 97–115. Göttingen: Vandenhoeck and Ruprecht.

——. 1983. The mythology of taxonomy. *Trans. Bot. Soc. Edinb.* 44:79–94.

——. 1984. Electronic data processing in taxonomy and systematics. In R. Allkin and F. A. Bisby, eds., *Databases in Systematics*, pp. 1–15. London: Academic Press.

——. 1985. Linnaeus: The conflict between science and scholasticism. In J. Weinstock, ed., *Contemporary Perspectives on Linnaeus*, pp. 1–16. Lanham, New York: University Press of America.

——. 1990. Botanic gardens and the conservation of plant resources. *Impact of Science on Society* 158:121–132.

——. 1993a. *Flowering Plants of the World.* London: Batsford; New York: Oxford University Press.

——. 1993b. The role of botanic gardens and arboreta in the ex situ conservation of wild plant resources. *Opera Botanica* 121:309–312.

Holttum, R. E. 1970. The historical significance of botanic gardens in S.E. Asia. *Taxon* 19:707–714.

Jacobs, M. 1980. Revolutions in plant description. *Misc. Papers Landbouwhogeschool Wageningen* 19:155–181.

Linnaeus, C. 1753. *Species Plantarum.* Holmiae.

Purseglove, J. W. 1959. History and function of botanic gardens with special reference to Singapore. *Gardens' Bulletin, Singapore* 17:125–154.

Raven, P. H. 1977. The systematics and evolution of higher plants. In *Changing Scenes in Natural Sciences, 1776–1976*, pp. 59–83. Philadelphia: Academy of Natural Sciences; Lancaster: [Spec. Publ. 12].

Raven, P. H. and E. O. Wilson. 1992. A fifty-year plan for biodiversity surveys. *Science* 258:1099–1100.

Robertson Vernhes, J. and T. Younès. 1993. Inventorying and monitoring biodiversity under the Diversitas programme. *Biology International*, no. 27, pp. 3–14.

Shetler, S. G. 1974. Demythologizing biological data banking. *Taxon* 23:71–100.

Stafleu, F. 1963. Adanson and his *Familles des plantes.* In G. H. M. Lawrence, ed., *Adanson: The Bicentennial of Michel Adanson's* Familles des plantes. pp. 123–264. Pittsburgh: Hunt Botanical Library.

Vanidiver, R. A. 1988. Integration of education with research and collections programs in North American botanic gardens. Masters thesis, University of Oklahoma, Norman.

WCMC (World Conservation Monitoring Centre). 1992. *Sampling Biodiversity: Status of the Earth's Living Resources.* London: Chapman and Hall.

Wilson, E. O. 1992. *The Diversity of Life.* London: Allen Lane.

WWF/IUCN/BGCS. 1989. *The Botanic Gardens Conservation Strategy.* Gland and Richmond: IUCN.

2

Societal and Scientific Information Needs from Plant Collections

BRIAN M. BOOM

The societal and scientific information needs from plant collections are often grossly underappreciated by the very individuals who either currently or potentially could use such information. This is unfortunate because of the tremendously important role collections play in providing baseline data about the extent and distribution of plant diversity. The utility of collections runs the gamut from the most basic, theoretical of scientific studies to the most applied, commercial of applications. This paper reviews the principal literature concerning the information needs of society and science that are met by systematic botany research and the collections that provide its underpinnings.

If the role of systematic botany in meeting the information needs of science and society is obscure to the general public, as most assuredly it is, then the role played by plant collections, the material basis for systematic botany, is even more elusive to all but the best informed. A general lack of understanding exists of the essential link between information needed by the scientific community and society at large and the potential of plant collections to provide that information. This is unfortunate for many reasons, but principally the result is that many opportunities are missed to improve human welfare and to make significant advances in our understanding of the natural world. Furthermore, much public and private money is wasted in ignorance of the things that plant collections could easily reveal. Never has it been so critical for decision makers to understand that their actions will work better, and often cost much less, if those actions have a rational, scientific basis. In terms of plant resources, that basis is achieved through various disciplines and techniques and at levels of organization ranging from the gene to the ecosystem, but when all is said and done, the underpinnings for all this activity are plant collections.

The purpose of this book is to explore some of the innovative concepts and techniques of collection, preservation, and storage of plant diversity. Before considering the actual collection, preservation, and storage of materials, it is appropriate to review the reasons for all this activity. What good are plant collections, anyway? In chapter 1, Vernon Heywood has presented a historical overview of the documentation of plant diversity. The purpose of the present paper is to review the myriad information needs that are, or could be, met by plant collections. Most of these needs are rather obvious to anyone who has thought about them, but the problem is that most people have never contemplated the issues, and the importance of plant collections has been taken for granted. If this paper serves no other purpose, I hope it will be to outline clearly, for scientists and nonscientists alike, the role of plant collections in fostering greater understanding of the natural world and in improving human welfare.

The Voucher Specimen

Before outlining the utility of plant collections, one concept must be clearly understood, namely, that of the voucher specimen. Simply put, a voucher specimen is one that serves as proof or supporting evidence for the identity of a particular species or infraspecific entity, and for the existence of that taxon at a certain location at a certain time. It is the "anchor" for all future reference to that taxon's occurrence, and it provides means for checking the identity in the future. In essence, the voucher specimen is the most basic element in plant science research.

No matter what the particular field of inquiry—ecology, genetics, systematics, physiology, ethnobotany, pathology, or biodiversity prospecting—if one does not collect a plant specimen and deposit it permanently in an herbarium, then there is no way for any work to be verified as to the identity of the organism involved.

The published literature is full of examples of mistakes that resulted because no voucher specimens had been made. Take the case of mistaken identity of the principal ingredient in the blowpipe dart poison of the Panare Indians in the Venezuelan Guayana. Anthropologist Paul Henley (1982) reported that *mankowa* was the Panare name of the plant that was made into curare, and he listed the botanical name as *Strychnos fendleri* Sprague & Sandw., a small tree. No voucher specimen was cited. In the course of my study of Panare ethnobotany (Boom 1990), I discovered that the real *mankowa* is *Strychnos toxifera* Schomb., a liana known widely throughout the Guayana region as a major ingredient in curares. My voucher specimen number cited was *Boom 6505*, and the specimens are deposited for permanent reference in the herbaria of The New York Botanical Garden and the Facultad de Farmacia, Universidad Central de Venezuela, the institution I was collaborating with on that particular project. There is no way these two species can be confused if specimens are collected, and I was thus able to correct the true identity of the Panare's dart poison.

In a few fields of biology, site records are sometimes considered adequate for certain purposes. In ornithology, for example, bird occurrence is often recorded on the basis of hearing calls or seeing the species in the field. But birds are much better known than plants and many fewer species of them exist. In botany, if there is no voucher specimen, one generally has no basis for knowing what taxon is being discussed. This would seem obvious to a practicing plant taxonomist, but regrettably it is often not yet so for workers in other disciplines, such as forestry, anthropology, and ecology, who often rely on local, indigenous names as the only basis for recording the occurrence of plant species. Sometimes such workers will attempt to match recorded local names with published lists of botanical equivalents for a given region. Such attempts result in unreliable published data.

Two examples of the unreliability of indigenous names come from my ethnobotanical study of the Chácobo Indians of Bolivia (Boom 1987). I recorded that the same local name *jihui coshi* (*jihui* = tree; *coshi* = strong) applied to two different species in different families: *Vochysia vismiifolia* Spruce ex Warming (Vochysiaceae) and *Diplotropis purpurea*

(Rich.) Amsh. (Leguminosae). Both species are highly valued for their strong wood, which is used in bridge construction. Likewise, sometimes different local names are given for the same botanical species. *Ischnosiphon arouma* (Aubl.) Koern. (Marantaceae) is called *manicoro* (*mani* = to join together; *coro* = mist), a leaf decoction of which is used to cure diarrhea and as an expectorant, and *xëco joni* (*xëco* = to embrace; *joni* = man), the main stem of which is used to make arrow shafts. A researcher who did not collect voucher specimens, and thereby had no way of properly identifying the species involved, would have generated major confusion in the literature.

Collections are the link between the plant in nature and the ultimate information to be derived from that plant. Collections are like libraries where information on plants is stored. These specimens, which form the base for systematic botanical studies, are every bit as priceless as libraries, and they deserve much more support and attention from the funding community than they currently receive.

The Specimen Base

In the United States there are 628 herbaria, containing a total of over 60 million specimens. Worldwide there are 2,639 herbaria, containing more than 272,800,000 specimens (Holmgren, Holmgren, and Barnett 1990). These collections constitute the primary database for the world's plant diversity.

The ongoing support of natural history collections is a subject of considerable concern among those charged with their stewardship. One attempt to raise funds for collections using their pricelessness as a selling point was undertaken by the Biology Curators Group at the Bolton Museum and Art Gallery in Great Britain. Their promotional brochure describing "The Sunflower Campaign: Biology Collections in Crisis" has an illustration of Van Gogh's painting *Sunflowers* with a price tag that reads 25 million pounds, the price the painting had been sold for in 1987, alongside a drawing of a Great Auk in a museum mount with the label Priceless. The point is made quite well. A principal consideration related to funding of the specimen base is the gridlock in space for new collections that most herbaria currently face. This lack of proper infrastructural support, together with the paucity of trained systematic botanists, has been termed the *second biodiversity crisis* (Boom 1991), the first crisis being the actual erosion of biotic diversity through habitat destruction, and so on, or the "taxonomic impediment" (Anonymous 1994:24). Unless the systematic community finds means to solve the second biodiversity crisis, there is little hope of being able to address the first crisis. For example, at my own institution, The New York Botanical Garden, we have the largest herbarium in the Western Hemisphere, with 5.7 million specimens. We also continue to add nearly 100,000 specimens to our collection each year. This has resulted in a gridlock space situation, and a new 60,000-square-foot Library/ Herbarium Building is being planned to solve the space problem. But that will not be completed until 1999, and meanwhile the specimens keep coming; our solution is to build a temporary, 3,000-square-foot herbarium building to handle incoming material until the new structure is finished. I do not want to dwell here on the topic of the second biodiversity crisis, but it is essential to understand in order to better appreciate the impediments that are often preventing the maximum use of plant collections.

The Value of Systematic Botany Collections

Various recent publications have enumerated the actual and potential values that biodiversity has to science and society. To the extent that systematics collections represent the base for what we know about biodiversity, these values may be extended to the collections themselves. Reid and Miller (1989) discussed the scientific basis for conserving biodiversity but focused principally on the role of biodiversity in ecosystems and scarcely mentioned the economic rationales.

The same year, the National Science Board published a report that summarized the dimensions of the biodiversity crisis, and highlighted the economic and social importance of biodiversity: "Human prosperity is based very largely on the ability to utilize biological diversity: to take advantage of the properties of plants, animals, fungi, and microorganisms as sources of food, clothing, medicine, and shelter" (National Science Board 1989:9–10). This document has yet to receive the full attention it deserves, but it does present an authoritative case for studying biodiversity, and, by extension, for the value of systematics collections.

However, the publications that concern systematics collections per se are of greater interest here, and it was during the late 1960s that several appeared. One contained proceedings of a symposium dealing with herbaria in modern universities (Beaman, Rollins, and Smith 1965).

Vernon Heywood's (1968) *Modern Methods in Plant Taxonomy* contained several pertinent chapters, namely, Brenan (1968) on the relevance of national herbaria to modern taxonomic research; Cronquist (1968) on the same topic, but restricted to the situation in the United States; and McNeill (1968) on regional and local herbaria. The following year Shetler (1969) presented an overview of herbaria: past, present, and future.

Certainly, among the most important earlier publications on the value of systematics collections was the so-called Steere Report (Conference of Directors of Systematic Collections 1971). This short but well-conceived document summed up the situation eloquently:

> The major systematic collections are man's treasures of information about his fellow creatures on earth. Today man is beginning to realize that his population and technological needs must be accommodated in a finite world. The challenge of the hour is for man to find ways—culturally, economically, aesthetically, and spiritually—to respond to this truth. The great systematic collections of the United States are the key to that understanding. If they are to fulfill their vital role, they urgently require the active support of the federal government. (Conference of Directors of Systematic Collections 1971:viii)

The survey was done for and supported by the National Science Foundation and was prepared by a committee of directors of the Conference of Directors of Systematics Collections (the forerunner of the Association of Systematics Collections).

In chapter 2 of the Steere Report (Conference of Directors of Systematic Collections 1971) ("Who Uses Systematic Biology Collections and How?"), which should be required reading for systematic biologists and students, several excellent examples are given of one of the realities of biology: "Because difference in kind reflects difference in structure, function, requirement, and relationship to the organic and inorganic world, the first order of business in any biological study is the identity of the organisms under investigation" (p. 5). Examples of world problems that are helped by systematic biologists in their use of plant collections include water pollution (algae), radiation, quinine, poisonous plant identification, and drug exploration. The role of systematics collections in the study of evolution and, indirectly, in the discovery of the very essence of the gene is cited. Also discussed is the large educational role played by systematics collections: "Only by studying, working with, and contributing to major systematics collections do biologists-in-training gain comprehensive perspective on the actual variety in today's biological world" (p. 8).

The next important publication was produced by a committee convened to write a report outlining a national program for systematics collections and their management (Irwin et al. 1973). This included a chapter on systematics in science and society, which lists applicable areas of plant collections such as health sciences, human food resources, the search for utilizable natural resources (minerals, energy, and construction materials), and societal concern for the environment (impact assessment, rare and endangered species, and indicator species of pollution).

One of the most cogent statements about the application of systematics to ecology in this report is that made by E. O. Wilson (1971:741):

> Most of the central problems of ecology today can be solved only by reference to details of organic diversity. Even the most cursory ecosystem analyses have to be based on sound taxonomy. And after the first broad measurements of energy flow and geochemical cycling have yielded their important but limited information, what remains of intellectual challenge stems chiefly from details of the biology of a particular species. The food nets, the fluctuation of population numbers and biomasses, the diel and seasonal rhythms, the rates and patterns of dispersal, the colonization of empty habitats, microevolution, physiological adaptation, and most other basic topics of ecology require a deep understanding of the biology of individual taxa. Progress depends not just on correct identification of species, but also on the mastery of larger taxonomic groups of the kind best achieved through deliberate specialization by taxonomists or taxonomically trained ecologists.

Irwin et al. (1973:7) also take up the relationship between education and systematics, listing four purposes that systematics collections serve in this regard: (1) identified specimens and related information for exhibits; (2) the basis for major natural history publications in great demand by the public; (3) an important resource for elementary, secondary, and postsecondary curricula in the life sciences; and (4) a tangible record of organic evolution and the diversity of nature.

Several years later The New York Botanical Garden published a report on systematic botany resources in the United States. This document (Payne et al. 1979), written by a committee of the American Society of Plant Taxonomists, focused on the cost of services provided by herbaria, but it also touched on the roles of herbaria in U.S. society. The summary of this section (Payne et al. 1979:18) bears quoting here:

> The resource collections of America are intricately integrated into all aspects of American society and business. They serve many needs for industry, for medicine, for military preparedness, for agriculture, for education, for science, and in the fight to improve the quality of life and to battle the forces of pollution and urban deterioration. They undergird the American enterprise with ready information and with staff capabilities that cannot presently be secured or guaranteed in any other way. Because the collections include plants and plant fragments accumulated from throughout the world and preserved since before the time America was settled by European immigrants, they are treasure houses both of data and of real bits of ecosystems and organisms now long gone from the wild.

Then, at the end of their report, Payne et al. (1979:80–84) provide an excellent list entitled "Possible Uses of Systematics Collections for Solution of Problems of Human Health, Food Resources, Environmental Quality, and Location and Utilization of Natural Resources," divided into several categories: public service; conservation, land-use planning, and recreation; public education; federal agency projects, including national defense; medicine and public health; agriculture and forestry; and business and industry.

Stuessy and Thomson (1981), in a report to the Systematic Biology Program of the National Science Foundation, assessed trends, priorities, and needs in systematic botany. Although not principally concerned with the utility of systematic botany collections, the publication contains a short, excellent section for basic systematic research entitled "Needs and Justification." The major value of this report, however, is the strong case it makes for the need to rectify what has been described above as the "second biodiversity crisis" or the "taxonomic impediment."

One of the most useful references for examples of the utility of systematics collections, both plant and animal, is the bibliography compiled by Knutson and Murphy (1988). Although the authors do not claim that the work is all-inclusive and that it is strongest in their own discipline, entomology, it is one of the best compilations of its kind. Particularly relevant sections include purposes and prospects for systematics; predictive capabilities of systematics; relationships of systematics with other fields (general, biological control, pest management, quarantine and regulatory activities, ecology, environment, plant germplasm, forestry, biomedicine, veterinary medicine, and genetics). Other relevant sections are on descriptions and analyses of collections resources for systematic work in botany, mycology, microbiology, and living collections. Rossman (1988:11) indicated that the bibliography will be maintained on a word processor, and updated versions will be made available periodically. The bibliography may be purchased for $9.50 from the Association of Systematics Collections, 1725 K Street NW, Suite 601, Washington, D.C. 20001, USA.

For several days in early May of 1988 a workshop entitled "Floristics for the Twenty-first Century" was held in Alexandria, Virginia, to discuss the current use and future plans for floristic information. More than sixty representatives from basic and applied research fields discussed this subject, primarily with a focus on vascular plants of North America north of Mexico. The results (Morin et al. 1989) contain extensive examples of how floristic information is used by science and society. Of special interest are several tables in Morin et al. (1989). Table 2 lists uses of floristic information in applied biology, under three categories: horticulture, crop development, and resource management. Table 3 lists the kinds of floristic information used in basic biological research. Finally, Table 5 enumerates forty-six potential uses of floristic information by nontaxonomists. Because floristic information is necessarily derived from systematic botany collections, this work is also an essential reference for anyone interested in this subject.

A report produced by the British House of Lords Select Committee on Science and Technology (1991) considers the state of systematic biology research in England. Chapter 3 of that thorough document concerns the uses of systematic biology research, focusing on economic applications: medicine, the pharmaceutical industry, agriculture, industrial fermentation and biotechnology, fisheries, customs control, biological recording, water pollution, and biological systematics as recreation.

Shortly thereafter, a report was published resulting from the Conservation and Preservation of Natural Science Collections Project, a joint effort of the National Institute for the Conservation of Cultural Property (NIC), the Association of Systematics Collections (ASC), and the Society for the Preservation of Natural History Collections (SPNHC). In this important document (Duckworth, Genoways, and Rose 1993) the first chapter is devoted to the significance and value of natural history collections, and the extensive bibliography at the end of the report has a category on the value and use of collections.

One of the newest and most promising sources of information about plants are gene banks. This emerging type of collection base was the subject of a symposium volume edited by Robert and Janice Adams (1992), entitled *Conservation of Plant Genes: DNA Banking and In Vitro Biotechnology*, based on the DNA Bank-Net Workshop held at the Royal Botanic Gardens-Kew in April 1991. I will not devote a great deal of time to this subject, because it is a focus of chapter 13. Nonetheless, because it is so important a topic, I do want to highlight several points.

One of the task forces at that Kew symposium was charged with projecting the uses of DNA from DNA Bank-Net. Several immediate uses for conserved DNA were summarized (Adams and Adams 1992:333):

1. Molecular phylogenetics and systematics of extant and extinct taxa;
2. Production of previously characterized secondary compounds in transgenic cell cultures;
3. Production of transgenic plants using genes from gene families;
4. *In vitro* expression and study of enzyme structure and function;
5. Genomic probes for research laboratories.

Obviously, there are tremendous scientific and societal information needs to be met by conserved DNA. There are, however, two dimensions to these potential uses as they relate to collections: (1) the issue of DNA samples that are expressly collected for long-term storage, as in liquid nitrogen, and (2) the use of existing herbarium and fossil plant material in natural history collections as a source of gene sequence data. Giannasi (1992) discusses the latter situation, in which he points out that some 5–20% of herbarium specimens sampled may be expected to possess usable DNA. This emerging new field has great possibilities, but the potentially widespread practice of extracting DNA from herbarium specimens holds concern for curators of these collections in that the process is somewhat destructive of the original material, especially in the case of plant fossils. Giannasi (1992:91) gives some guidelines for the use of herbarium/fossil DNA that curators will want to consider for the collections under their care.

The most recent, important document to highlight the values of systematic biology—and, by extension, the collections on which the discipline is based—is *Systematics Agenda 2000: Charting the Biosphere* (Anonymous 1994). Actually consisting of two versions—a short, nontechnical one with attractive color photographs and graphics, and a more lengthy, technical version—this publication represents a landmark achievement involving the collaboration of more than three hundred systematists across the United States. *Systematics Agenda 2000* outlines a global initiative to discover, describe, and classify the world's species over the next twenty-five years. Both technical and nontechnical documents contain a variety of excellent examples of the utility of systematics collections to science and society, and they deserve wide circulation to decision makers in government, industry, and academe. Copies of *Systematics Agenda 2000* may be obtained by writing to SA2000, Herbarium, The New York Botanical Garden, Bronx, New York 10458, USA.

Literature Cited

Adams, R. P. and J. E. Adams, eds. 1992. *Conservation of Plant Genes: DNA Banking and In Vitro Biotechnology*. San Diego: Academic Press.

Anonymous. 1994. *Systematics Agenda 2000: Charting the Biosphere*. New York: Systematics Agenda 2000.

Beaman, J. H., R. C. Rollins, and A. H. Smith. 1965. The herbarium in the modern university: A symposium. *Taxon* 14:113–133.

Boom, B. M. 1987. Ethnobotany of the Chácobo Indians, Beni, Bolivia. *Adv. Econ. Bot.* 4:1–68.

———. 1990. Useful plants of the Panare Indians of the Venezuelan Guayana. *Adv. Econ. Bot.* 8:57–76.

———. 1991. Biological diversity crisis II. *Association of Systematics Collections Newsletter* 19 (5): 63–64.

Brenan, J.P.M. 1968. The relevance of the national herbaria to modern taxonomic research. In V. H. Heywood, ed., *Modern Methods in Plant Taxonomy*, pp. 23–32. London: Bot. Soc. Brit. Isles/Linn. Soc. Lond./Academic Press.

Conference of Directors of Systematic Collections. 1971. *The Systematic Biology Collections of the United States: An Essential Resource*. Part 1: *The Great Collections: Their Nature, Importance, Condition, and Future*. Bronx, N.Y.: The New York Botanical Garden.

Cronquist, A. 1968. The relevance of the national herbaria to modern taxonomic research in the United States of America. In V. H. Heywood, ed., *Modern Methods in Plant Taxonomy*, pp. 15–22. London: Bot. Soc. Brit. Isles/Linn. Soc. Lond./Academic Press.

Duckworth, W. D., H. H. Genoways, and C. L. Rose. 1993. *Preserving Natural Science Collections: Chronicle of Our Environmental Heritage*. Washington, D.C.: National Institute for the Conservation of Cultural Property.

Giannasi, D. E. 1992. Feasibility of obtaining comparative gene sequence data from preserved and fossil materials. In R. P. Adams and J. E. Adams, eds., *Conservation of Plant Genes: DNA Banking and In Vitro Biotechnology*, pp. 75–98. San Diego: Academic Press.

Henley, P. 1982. *The Panare: Tradition and Change on the Amazonian Frontier*. New Haven: Yale University Press.

Heywood, V. H., ed. 1968. *Modern Methods in Plant Taxonomy.* London: Bot. Soc. Brit. Isles/Linn. Soc. Lond./Academic Press.

Holmgren, P. K., N. H. Holmgren, and L. C. Barnett. 1990. *Index Herbariorum.* Part 1: *The Herbaria of the World.* 8th ed., *Regnum Vegetabile,* vol. 120. New York: The New York Botanical Garden.

Irwin, H. S., W. W. Payne, D. M. Bates, and P. S. Humphrey, eds. 1973. *America's Systematics Collections: A National Plan.* Lawrence, Kans.: Association of Systematics Collections.

Knutson, L. and W. L. Murphy. 1988. *Systematics: Relevance, Resources, Services, and Management: A Bibliography.* Washington, D.C.: Association of Systematics Collections.

McNeill, J. 1968. Regional and local herbaria. In V. H. Heywood, ed., *Modern Methods in Plant Taxonomy.* London: Bot. Soc. Brit. Isles/Linn. Soc. Lond./Academic Press.

Morin, N. R., R. D. Whetstone, D. Wilken, and K. L. Tomlinson. 1989. *Floristics for the 21st Century. Monographs in Systematic Botany* 28. St. Louis: Missouri Bot. Gard.

National Science Board. 1989. *Loss of Biological Diversity: A Global Crisis Requiring International Solutions.* Washington, D.C.: National Science Board.

Payne, W. W., T. B. Croat, T. S. Elias, P. K. Holmgren, R. McVaugh, D. H. Nicolson, L. I. Nevling, Jr., R. Ornduff, and R. F. Thorne. 1979. *Systematic Botany Resources in America.* Part 2: *The Cost of Services.* Millbrook, N.Y.: Cary Arboretum of The New York Botanical Garden.

Reid, W. V. and K. R. Miller. 1989. *Keeping Options Alive: The Scientific Basis for Conserving Biodiversity.* Washington, D.C.: World Resources Institute.

Rossman, A. Y. 1988. Editor's note: Systematics: Relevance, resources, services, and management. *Association of Systematics Collections Newsletter* 16 (2): 1–12.

Select Committee on Science and Technology. 1991. *Systematic Biology Research.* Vol. 1: *Report.* House of Lords Session 1991–92, 1st Report, HL Paper 22–1. London: HMSO.

Shetler, S. G. 1969. The herbarium: Past, present and future. *Proc. Biol. Soc. Wash.* 82:687–758.

Stuessy, T. F. and K. S. Thomson, eds. 1981. *Trends, Priorities, and Needs in Systematic Biology.* 2d ed. Lawrence, Kans.: Association of Systematics Collections.

Wilson, E. O. 1971. The plight of taxonomy. *Ecology* 52:741.

3

The Urgency of Documenting Plant Diversity: The Flora of the Philippines Project—A Last Chance to Study the Plant Diversity of the Philippines

S. H. SOHMER

The Philippine Flora Project was conceived as a response to the rapid rates of deforestation in the Philippines that threaten to eliminate the majority of endemic plants of the archipelago. During fieldwork in the Philippines in the early 1980s, entire forests were seen to be disappearing. It was clear that this flora was on the verge of total destruction and that not very many people in the country could or would do anything about it. Dr. Domingo Madulid, of the Philippine National Museum, suggested the concept of a new Flora of the Philippines. This project is documenting the remnant flora of the archipelago through a rigorous and systematic collecting and exploring program, strengthening the botanical infrastructure of the Philippines, and preparing the ground for the first comprehensive flora of the country. Deforestation in the Philippines is the result of a high rate of population growth (ca. 2.8% at present) and concomitant exploitation of the country's natural resources, fueled by foreign interests in partnership with powerful, influential, and wealthy individuals within the country who often allow little opportunity for indigenous groups to participate in determining the fate of traditional lands. Unrestricted, massive logging opens the land to occupation by homeless and hungry peasants who, within a few years of kaingin (slash and burn) activity, reduce the ability of the land to support anything other than cogon *grass* (Miscanthus *spp.) and other species (e.g.,* Imperata cylindrica, Themeda *spp.).*

The Philippine Flora Project (PFP) is a good example of research collaboration. The project originated in the collaborative interaction between myself and my colleague, Dr. Domingo Madulid, in the early 1980s centered on fieldwork with the genus *Psychotria* (Rubiaceae) in the Philippines. Dr. Madulid had wanted to initiate a Flora of the Philippines project, and I had become much concerned about the rapid disappearance of indigenous forests there. This led to our planning a strategy for documenting the remaining flora, including an understanding of its current status, strengthening the botanical infrastructure in the Philippines, and heightening public understanding and awareness of the native vegetation within the archipelago. It was within this conceptual framework that the project began, supported by the National Science Foundation and the MacArthur Foundation. It was initially a collaboration between the Bishop Museum in Honolulu and the Philippine National Museum in Manila, but it shortly thereafter became a collaborative effort between the Botanical Research Institute of Texas (BRIT) and the Philippine National Museum when U.S. personnel relocated to BRIT.

The Philippine Archipelago

Geography and Geology

The Philippine archipelago is in the western Pacific, occupying the region between the Pacific and Asian Plates roughly 3–21° North and 117–127° East. The western edge of the Pacific Plate is forced and subducted beneath the Asian Plate by the activity of mid-oceanic rifts. This constant abrasion explains the high seismic and volcanic activity throughout the archipelago, and it is nearly a mirror image of what occurs on the other side of the Pacific Ocean on the western coast of North America. Volcanic activity and coral reef uplift, and other forms of geotectonic activity have created considerable geological diversity in the Philippines that has influenced the development of high levels of plant and animal diversity.

There are about 7,100 islands in the Philippines, ranging in size from very large, such as Luzon, to barely peeping over the surface of the sea. The vast majority of the islands are small and the entire archipelago can be viewed as consisting of three or four parts: (1) Luzon and closely related islands, such as Mindoro; (2) the Visayas, which are the broad sweep of islands in the middle of the archipelago with large islands such as Negros and Samar; and (3) Mindanao (and its outliers, such as Basilan). Often Palawan is conceptually placed in this latter group, but its flora is more clearly related to that of Borneo because Palawan and Borneo were connected to each other by dry land until the recent geological past; thus Palawan can be placed in a biogeographical province within the Philippines by itself (fig. 3.1).

Biogeography

The flora of the Philippines is part of the great Malesian flora. It is a piece of a natural phytogeographic zone that includes Peninsular Malaysia, Sabah, Sarawak, Brunei, Indonesia, Papua New Guinea, and the Philippines (fig. 3.1). The Malesian phytogeographic zone probably contains at least thirty thousand species, many of which have not yet been described, and probably more than twenty-five hundred genera. It is one of the richest, as well as one of the most important centers of plant diversity in the entire world (Van Steenis, 1948a). The boundaries of the Malesian phytogeographic area are the isthmus of Kra between Peninsular Malaysia and Thailand, the Torres Straits between the island of New Guinea (including the Louisiades and the Bismarck Archipelago) and Australia, and the area between the Philippines and Formosa (Van Steenis 1948b). The flora of the Philippines contrasts starkly with that of Formosa, which is essentially a Sino-Japanese flora.

There has been movement and migration of plants between these floristic zones. There are examples of elements of the Sino-Japanese flora that migrated into the Malesian flora in the Philippines. Van Steenis (1948b) points out the example of *Lilium philippinense* Baker, the only native Philippine representative of a genus with about seventy species in the Northern Hemisphere and about thirteen species in the Old World (including the Philippine species already mentioned). Of

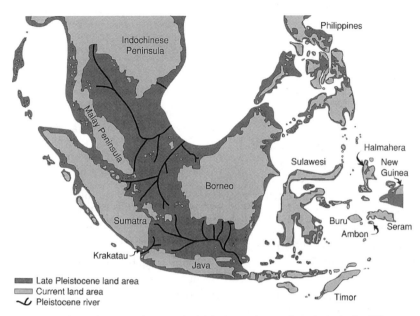

FIGURE 3.1   An outline map showing the Malesian region, and particularly the difference between the present islands that are part of the Sunda Shelf and those that are not. Redrawn from Heany, *Bull. American Mus. of Nat. Hist.* 206 (1991): 147.

course examples of the reverse also exist, with species from genera centered in the Malesian floristic zone migrating into neighboring zones, such as Japan. However, more than half the genera, and most of the species, found in the Malesian area do not occur in neighboring regions (Van Balgooy 1971).

Van Steenis (1948b) recognized three principal subdivisions of the Malesian flora based on distribution patterns of the endemic genera of seed plants. This has been further expanded by van Balgooy (1971). The generic phytogeographic relationships of the Philippines are with Borneo, Sumatra, and Peninsular Malaysia. The other two regions comprise, respectively, New Guinea, Suluwesi (the Celebes) and associated islands, and Java and the smaller islands to the East. About forty genera of seed plants exist that are endemic to the Philippines (Van Steenis 1948a). Obviously the more information we have on the distribution of plants, the more data we have for insight into past plant migrations. Climate and plate tectonics, particularly the latter, have probably had the most to do with the distribution of the current living plants of the world, including the overall pattern of plants in the Malesian area.

Species Endemism

The number of species endemic to the Philippines is not known with any degree of accuracy, but various estimates have been made. Merrill (1923–26) thought that overall species endemism may be as high as 75%, but most botanists think that the percentage of species endemism is probably closer to 25% (Madulid, pers. comm.). Certainly, much of the flora of the Philippines appears to be shared with neighboring areas, most particularly Borneo. As mentioned above, until rather recently in geologic time, there were direct connections between the present islands of Palawan and Borneo. Palawan is, like Borneo, part of the Sunda Shelf (fig. 3.1). When sea levels were much lower, the distances between the Bornean-Palawan land areas and the rest of the Philippines, particularly Mindanao, were much less than they are today. Migration of biota would have been greatly facilitated, except for the largest animals and plants with the heaviest fruits and seeds.

The Case for "Local" Floras

That the flora of the Philippines is a part of the great Malesian phyto-geographic zone lends weight to the arguments developed primarily by Dutch botanists, most notably van Steenis, that there should not be attempts at producing "local" floras until a comprehensive regional one is completed. These arguments stress that local floras tend to contain disparate treatments without a uniform taxonomic approach. Van Steenis (1948a) states in the introduction to his volume 4, part 1 of *Flora Malesiana* that "only temporary profit may be gained from making local floras, and both valuable time and money are wasted by the enormous duplication which is unavoidable when the goal of a flora of a plant-geographical unit is to be reached along this tortuous road."

The only problem with this logic is that tremendous destruction of the natural ecosystems has occurred, and is occurring, since those words were written; a level of destruction unforeseen at the time. Destruction is the single most potent argument for undertaking and completing as many local floras as possible at this time, which can serve as rallying points for the conservation of remnant natural areas. I agree that were it not for the fact that all the remaining natural areas in the Malesian floristic region are being literally destroyed before our eyes, there would be less cogent scientific argument against the comprehensive regional floristic treatment to which van Steenis had so ardently and fiercely dedicated his life. However, even in that case, there might still be potent political arguments for local floras, because they can more effectively mobilize political, social, and financial support than can a supranational effort. [Good examples include the *Tree Flora of Malaya* (Whitmore 1972) and the on-going project to prepare a similar Tree Flora of Sabah and Sarawak (Wong, Khoon-meng, Sabah Forest Research Centre, pers. comm.)].

In any case, we certainly have no time for argument now. The Flora of the Philippines Project will leave in its wake a much better understanding of the plant life of that archipelago than would otherwise have been possible. Population growth, political constraints, and funding realities have conspired to make the arguments for a regional flora advanced by van Steenis and his colleagues an ideal to which we may aspire but in the meantime must approach in a more diverse and pragmatic fashion.

Population Growth and Trends

The human population of the Philippines has increased enormously over the past fifty to ninety years. The first census undertaken near the beginning of this century, after the U.S. government was installed, indicates that about 5 million persons were living in the Philippines at that time. Several censuses in the late 1980s, including one carried out by the World Bank, indicated that there were more than 61 million people in 1987. Unofficially, the population at the present time may be as high as 80 million but is officially cited as about 65 million. If these data are correct, then the population has increased sixteen times in less than ninety years. The rate of increase at present is believed to be about 2.8% per year, although official government figures give 2.3 or 2.4%. The lower official government rate will result in a population of some 125 million in the year 2020, and the higher in a population of some 140–145 million by 2020. This population level will surely destroy whatever is left of the country's natural vegetation. The relentless population growth is the primary cause of the destruction of native ecosystems, but destructive logging and farming practices independent of the population growth has greatly contributed to the situation.

Phytodiversity and Potential Rates of Extinction

Forest Destruction

The original forest covered over 94% of the Philippines (Sohmer 1989). This had been reduced to less than 6% by 1989 (Sohmer 1989) and, judging by reports of field teams working on the Flora of the Philippines project (Stone 1992), there is probably less than 4% remaining. Even now, representation of the remaining forests of the Philippines is usually illustrated as great interconnected blocks centering on the central mountainous backbone of the islands on maps reproduced in various papers, such as the one produced by Tan and Rojo (1988), but the actual situation is more like that shown in figure 3.2 (Sohmer 1989). These forest remnants are scattered but nevertheless hold the key to potential regeneration of biodiversity of the Philippine archipelago in the future. All efforts to save what is left should be made now. The felling rate reported by Petocz (1989) was 170,000 hectares per year, and this probably still holds true today. After felling, the former forested area is usually invaded by *kaingineros*, slash and burn agriculturists, who deplete the fertility of the land within a few years and then have to move on. It is unlikely that these forested areas will ever be regenerated again by native species, although weedy, secondary jungles and quasi-monocultures of "reforestation" species of various genera (*Acacia, Gmelina, Leucaena, Pterocarpus,* and *Swietenia*) may develop.

Numerous conservation efforts are in fact being planned or implemented at this time by a collection of national and international aid and conservation organizations. Most activity centers on the Integrated Protected Management System, with management of some parks potentially handed over to nongovernment conservation or environmental organizations by an overworked, underpaid, and understaffed Parks and Wildlife Department. Perhaps these efforts will save what is left of Philippine biodiversity or at least slow down the rates of extinction.

FIGURE 3.2   The remaining old growth forests of the Philippines are now often small, isolated, and scattered remnants, as shown in this figure (Sohmer 1989). Map adapted and redrawn from data produced for the Government of the Philippines Master Plan for Forestry Development showing remnant forest areas in the Philippines as of late 1988.

Probable Rates of Extinction

Of the original flora of the Philippines, about 8,000 species, as discussed above, it is estimated that a significant, but presently unknown number are endemic to the archipelago. The greatest number of these endemic species are found in native forests. When these forests are cut, obviously, the species endemic to them are extirpated. I hypothesize that, by the turn of the century, it is possible that as much as 90% of the original endemic forest flora will be gone. An example of rates of extinction can be seen in the statistics associated with the collecting history of *Psychotria* in the Philippines. Sohmer (in press) recognizes 127 species in this genus indigenous to the archipelago. About 42% of these species are represented only by collections made before 1930. After 1930, despite an increase in collecting activity after the war and the heavy collecting by Sohmer in the 1980s, these species have not been re-collected.

The Philippine Flora Project

The Philippine Flora Project was designed to accomplish three objectives: (1) document the remaining flora as quickly as possible through a countrywide collecting effort; (2) strengthen the botanical infrastructure of the country by helping to upgrade the facilities at the National Herbarium; and (3) produce a new comprehensive accounting of the flora of the Philippines that will provide a contemporary understanding of the native and naturalized plants of the archipelago. To these original objectives was subsequently added a fourth: the "rescue" of endangered species via ex-situ conservation.

The organization of the Philippine Flora Project consists of a Steering Committee of staff members of the Botanical Research Institute of Texas (Fort Worth) and Philippine National Museum (Manila). This committee has overseen and guided the project's active components—the Philippine Plant Inventory (PPI) and the Collections Management Project (CMP)—and has planned for the third part of the project—the Research/Editing Component. The PPI was initiated with funding from the U.S. National Science Foundation (NSF) in 1990, and the CMP was initiated with funding from the MacArthur Foundation in 1991. A new grant from NSF will continue the PPI until 1997, as it has been determined that the most timely component of this project is the collecting and documentation of the flora because of the rapid destruction of remaining Philippine natural areas. The first session of a workshop to help plan the rest of the project was held on 26–28 May 1994 in Fort Worth, and the second session was held in Manila from 29 September to 1 October.

The Philippine Plant Inventory

The concept of the PPI is a six-year effort divided into two phases. The first phase was completed in 1993, and the second phase began early in 1994. The PPI is designed to organize the exploration, collection, and specimen-processing activities that will produce the material and observational evidence on which our current and future knowledge of the Philippine flora will be based. In addition to organizing the field-work by three field teams whose main goal is to cover every part of the archipelago in that six-year period at least twice (in different seasons), the PPI dictates the establishment of three semipermanent field plots to be visited at least semiannually for intensive collecting and observations. In this way we will better assess how fieldwork has covered the archipelago during regular fieldwork. For example, if intensive collecting twice a year in semipermanent plots uncovers 40% more taxa on average per unit area than regular survey sites, we may conclude that site collections will not be fully representative of the total diversity present. By extrapolation, we will better understand the potential representation of the collections.

Collecting teams of the PPI are in the field more than half the year. Each team usually completes about six expeditions per year, each lasting three to five weeks, and each is followed by a period of about two and one-half weeks in the Philippine National Herbarium (PNH) at the National Museum working on their collections. Each team leader is experienced and educated (with a bachelor of science degree or higher in botany or forestry), and the team members generally have forestry or biological backgrounds and practical, on-the-job experience.

Collecting is intended to be opportunistic but representative, and where at all possible, comprehensive. Additional assistance is obtained by short-term hire of local guides, laborers, and tree-climbers, which facilitates exploration and produces a beneficial rapport with the local people. Two vehicles are available to the PPI, each able to transport at least six persons and several hundred pounds of cargo. Most areas on Luzon are accessible by road, and by use of ferries one can proceed to several Visayan islands and even to Mindanao. Very small islands and the more remote western and southern islands are reached by sea where local transport (jeepney) is available for hire. Local conditions such as weather and security are monitored, and if conditions are adverse, the trips are rescheduled or substituted. Within a specific area, collecting proceeds until it is evident that the area has been worked out for that particular trip.

The responsibilities of the collecting teams when at the museum include drying, labeling, filing, identifying specimens, and filing trip reports. Since the alcohol method is used to prevent specimen deterioration in the field, specimen papers need to be changed and paper corrugates added when the specimens are placed in presses and dried. All specimens are tagged with a PPI collection number.

Collaborating botanists join the collection teams to explore and collect their special groups. The vehicle and the general supplies are provided by the project, and the collaborators supply their own cost of reaching the Philippines and their own room and board. Collaborators are required to provide sets of their collections to the project.

Results of PPI

We have found that a team can obtain 2,000–3,000 specimens per expedition (i.e., about 300–400 collection numbers with about 7 to 10 replicates each). During 1991–92, for example, 9,031 collections were made with a total of about 80,000 specimens. These numbers will decrease as we learn which species are being "overcollected" and as the remaining unvisited sites are less accessible. The hope is that the PPI will fully cover the remnant vegetation of the country and thereby provide what will essentially be the last view of what is left of the flora. The specimens collected, along with specimens from historical collections that reside in the world's herbaria, will provide the basis of the data that will be needed for writing a new comprehensive flora of the Philippines.

Collection Management Project

The National Museum of the Philippines has been much neglected over many decades. Despite the fact that the Supreme Court and the Senate occupy the same building, the former Executive House Building of the

American colonial government, there has been little benefit to the museum. The museum generally has the lowest governmental priority for support, and the National Herbarium has shared in that general lack of support despite the recent significant increase in bilateral and multilateral support from foreign aid organizations for conservation activities, some of which has come the museum's way. Unfortunately the economic situation of the Philippines is grave and was considered not likely to improve significantly. It was therefore unrealistic to expect great improvements in the condition of the herbarium from Philippine governmental sources. For this reason, infrastructure strengthening of the herbarium, in support of the Flora Project, was designed into the project from the start. The principal goals are to reduce the backlog of unprocessed material of some thirty thousand collections, upgrade the physical condition of existing collections, and improve the general work and storage environment (especially by air-conditioning). The late Dr. Benjamin C. Stone, Senior Research Botanist at BRIT, had worked particularly on this aspect of the project supported by the MacArthur Foundation.

Research/Editing Component

This component of the project, in which the actual volumes providing a detailed and comprehensive accounting of the flora will be produced, is not yet operational. It will be the most difficult and complex of the project's three components. The intention now, based on the recent workshop, is to form "core" groups of salaried workers at BRIT and the National Herbarium in Manila, plus paid staff at supporting institutions and volunteer collaborators all over the globe. It may require as many as eight written volumes for this flora if illustrated, or five to six if not. The project design provides for the training of Philippine students and botanists. The Philippine institutions must also support the intellectual environment that must be part of this endeavor and contribute in some meaningful way to the efforts. In all, we are well aware that we are engaged in a monumental task.

As mentioned above, in May 1994 the first session of the NSF-funded Philippine Flora Workshop was held in Fort Worth. The workshop focused on the research/editing component. In addition to the printed flora, a computerized database will be developed consisting of all floristic and taxonomic information available for the native and naturalized plants of the Philippines. The database, updated and maintained, will provide a means to revise the Philippine flora. The Botanical Research Institute of Texas and the Philippine National Museum will be the two principal organizations along with supporting institutions, the Smithsonian Institution (Washington, D.C.), Rijksherbarium (Leiden), and the Royal Botanic Gardens, Kew. The principal research effort will be centered at the Philippine National Museum, and the principal editing effort, as well as primary responsibility for maintaining the database, will be centered at the Botanical Research Institute of Texas.

Acknowledgments

I thank my long-time colleague and friend Domingo Madulid, without whom this project could not have achieved its present reality. I pay tribute to the late Ben Stone's wisdom, leadership, and collaborative spirit, which have had a great deal to do with establishing the project over which he exercised such kindly, wise, and benign control. To the many individuals in the Philippines who work for the project or who, like Father Gabriel Casal—the Director of the National Museum—and Senator Heherson Alvarez, have helped the project from the "top," I extend warm regards and many thanks.

Literature Cited

Merrill, E. D. 1923–26. *An Enumeration of Philippine Flowering Plants*. Manila: Bureau of Printing.

Petocz, R. 1989. *Establishment and Management of an Integrated Protected Area System*. Draft report. Manila: World Bank.

Sohmer, S. H. 1989. *Preservation and Maintenance of Biological Diversity*. Annex D to Natural Resources Profile of the Philippines. Manila: U.S. AID.

——. In press. The genus *Psychotria* (Rubiaceae) in the Philippine Archipelago. *Sida, Bot. Misc.*

Stone, B. C. 1992. January 1992 marks first anniversary of PPI expeditions. *The Philippine Flora Newsletter* 2:1.

Tan, B. C. and J. P. Rojo. 1988. The Philippines. In D. C. Campbell and M. D. Hammond, eds., *Floristic Inventory of Tropical Countries: The Status of Plant Systematics, Collections, and Vegetation, plus Recommendations for the Future*, pp. 46–62. New York: The New York Botanical Garden.

Van Balgooy, M. 1971. Plant geography of the Pacific. *Blumea* 6:1–222.

Van Steenis, C. G. G. J. 1948a. Introduction. In C. G. G. J. van Steenis, ed., *Flora Malesiana*, series 1, vol. 4, pt. 1. Batavia (Jakarta): Noordhoff-Kolff N. V.

——. 1948b. General considerations. In C. G. G. J. van Steenis, ed., *Flora Malesiana*, series 1, vol. 4, pt. 4. Batavia (Jakarta): Noordhoff-Kolff N. V.

Whitmore, T. C., ed. 1972. *Tree Flora of Malaya: A Manual for Foresters*. London: Longman.

PART 2

Collection of Plant Materials

Before plant materials can be preserved and stored for generations, they must be collected. Bernard Baum in chapter 4 points out that there are 2,639 herbaria in the world containing approximately 270 million specimens. He also lists 1,300 botanic gardens and 1,023 gene banks, both of which maintain numerous living and genetic plant resources. With approximately 300,000 species of plants on earth, this sample of nearly 1,000 sheets per species might seem adequate for our needs. Baum, however, shows convincingly that this sample is not statistically adequate, nor is it likely ever to be so, especially in the face of diminishing diversity through human population increase and corresponding alteration of the environment. The only conclusion possible, therefore, is that we must attempt to set aside as many native areas and preserves as possible—areas in which we can in the future carry out much larger samples of selected taxa, depending on the specific questions under study. Conservation of taxa is important, but for really increased samples of the natural world, we will have to depend on preservation of taxa in natural areas and reserves.

At the present time throughout the world there is much activity in screening for new pharmaceuticals, especially in poorly inventoried tropical regions. The ravages of disease in the human population, especially resulting from cancer and AIDS, have occasioned broad screening of native plants for possible antitumor, antiviral, and other medicinals. The successful use of taxol, from the evergreen *Taxus baccata*, shown to inhibit the growth and spread of ovarian tumors, is a well-known case in point. Recent extracts from *Ancistrocladus korupensis* (Ancistrocladaceae) of Tanzania, which shows promise in fighting the AIDS virus, is another conspicuous example. James Miller in chapter 5 gives many of these case studies, and shows appropriate techniques for sampling plant materials in bulk for successful extraction and analysis of active natural products. The best conditions for preserving samples for pharmaceutical research are on open racks, in a dry room, with warm (not hot) forced air. The importance of voucher specimens is also stressed. Once material is pulverized and extracted, the only way to check identifications is with an affiliated, well-preserved herbarium sheet deposited in a reliable institution.

4

Statistical Adequacy of Plant Collections

BERNARD R. BAUM

Habitat destruction with the resulting loss of genetic diversity and species extinction, especially in the tropics, is increasingly being perceived as a crisis of unforeseen consequences to humankind. For at least the last two decades it has been known that the pool of genetic diversity used to improve our crops, drawn from primitive varieties, has already been largely eroded. The genetic diversity of wild relatives of our crops, increasingly used as sources of genes for crop improvement, is also eroding, primarily because of overgrazing or habitat destruction.

The main sources of knowledge of biodiversity are the study of specimens acquired through exploration and their resulting collections. The status of collections is first examined from a historical perspective. Then collections of herbaria, gene banks of cultivated crops and their wild relatives, and botanical gardens are summarized. Sample size required for studying diversity and measures of diversity are briefly described. Clearly, from a number of reasons discussed, collections are deficient. This is because of inadequate support for systematics.

> As stewards of the planet's biota, the task of conservation
> biologists, for the time being, is to prevent extinction rather than
> create new life forms.
> —O. H. Frankel and M. E. Soulé, *Conservation and Evolution*

Life on this planet started at least 3.5 billion years ago. During that time a number of major diversification events took place, followed by a series of mass extinctions (Raup and Sepauski 1984). According to current knowledge, eukaryotes appeared at least 1.5 billion years ago and marine multicellular organisms appeared 700 million years ago. About 430 million years ago life invaded the land for the first time. The first representatives of the groups that are now dominant appeared on land: vertebrate animals, arthropods, fungi, and plants. The trend of biological diversity has ostensibly been increasing during the last 100 million years with the rapid radiation of angiosperms (Crepet, Friis, and Nixon 1991). Despite this recent trend in the history of life on this planet, it seems that 99% of the species that existed at one time or another have become extinct (Raup 1981).

So, what biodiversity crisis? Extinction is a natural phenomenon. Of the few theories, a highly speculative one argues for a periodic mass extinction approximately every 26 million years (Raup and Sepauski 1984), which may have coincided with meteorite impacts. We know that the last major extinction event, which led to the disappearance of about two-thirds of the terrestrial organisms at the time, occurred at the end of the Cretaceous Period 65 million years ago. On a smaller timescale it has been suggested that tropical rain forests were reduced to small refugia during the Pleistocene glaciations (e.g., Simberloff 1986). On a still smaller scale it has been argued that a speciation event in angiosperms occurs every ten to fifty years. It has also been argued that a natural (nonanthropogenic) extinction event (of a species) may occur every four months to once every three years (Raup 1981). The study of the extinction process is difficult to pursue, but it is essential in order to understand the impact of the perceived biodiversity crisis. Hey's (1992) attempt to develop a model based on cladograms is a good example of the complexities involved in estimating speciation events over time.

At issue here is time, specifically our times as opposed to the geological timescale. The exact causes of anthropogenic loss of biological resources are not known. It is now recognized, however, that they have been lost through large-scale clearing and burning of forests since the

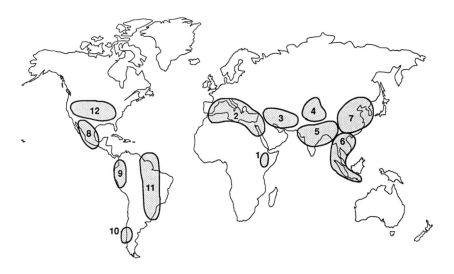

FIGURE 4.1 Centers of diversity of cultivated plants after Darlington and Janaki Ammal (1945). 1, Ethiopia; 2, Mediterranean coast; 3, Caucasus—Eastern Turkey and Iran; 4, Afghanistan; 5, Indo-Burma; 6, Siam—Malaya—Java; 7, China; 8, Mexico; 9, Peru; 10, Chile; 11, Brazil—Paraguay; 12, the United States.

beginning of agriculture and at an increasingly alarming rate in the last decade, by the overharvesting of plants and animals, the conversion of wildland to urban uses, and through other overexploitative activities and pollution (McNeely et al. 1990). This massive environmental destruction is a worldwide phenomenon that reduces, at an unprecedented historical scale, not only the number of species but the amount of genetic variation within individual species (Wilson 1985), especially in tropical countries [see reports in Campbell and Hammond (1989)].

The first ones to recognize this alarming situation were crop geneticists and taxonomists witnessing a rapid genetic erosion of landraces of cultivated crops and their wild related species. The landraces are early or traditionally used varieties that are highly variable genetically in contrast to our modern varieties that are genetically uniform. They called for the collection and preservation of crop genetic resources, including the wild related gene pools (see Frankel and Bennett 1970). The importance of wild species in cultivated crop evolution and breeding for crop improvement became recognized through cytogenetic studies. These led to an understanding of the processes and the significance of natural gene transfers from landraces and wild species to cultivated crops. Awareness of the danger of extinctions in geographic centers of diversity of crops and their wild relatives (fig. 4.1), upon which we depend so much, was raised as a critical issue as early as three to four decades ago (Harlan 1961). Recently, the disappearance of mushrooms in Europe is becoming a serious issue (Cherfas 1991), especially in connection with the possible disastrous effect on forest trees that depend on them because of their symbiotic association. Humankind recently became aware of the crisis through the knowledge generated over the years by plant scientists, especially systematists and geneticists.

Extinction is a failure of a taxon to perpetuate itself. The factors contributing to extinction are, as classified by Frankel and Soulé (1981) biotic (competition, predation, parasitism, and disease); isolation; and habitat alteration (slow geological change, climate, catastrophe, and humans). A number of approaches have been created to salvage what remains or to prevent the process leading to extinction caused by the human factor. In this regard it is useful to distinguish between preservation and conservation following Frankel and Soulé (1981). Preservation provides for the maintenance of individuals or groups, whereas conservation includes the policies and programs for long-term retention of ecosystems to provide the potential for continuing evolution.

Collecting specimens is a basic and necessary undertaking for any research about the earth's biota for utilitarian purposes as well as for the enrichment of our knowledge, which in turn changes our perception about the world and shapes our values. The purpose of this chapter is to examine some aspects of collections with emphasis on their statistics.

Historical Perspective

For the Greeks and Europeans during the Middle Ages, the study of medicinal plants was renowned as reflected in the herbals, the most famous being Theophrastus, Dioscorides, Brunfels, Bock, and Fuchs. In many of these herbals the plants were arranged according to their practical use. Then, the long voyages by European explorers, such as Marco Polo to the Far East and Columbus to the Americas, led to the discovery and recognition of biodiversity on the globe. It is noteworthy that these voyages were financed by royalty who had an interest in new economic species of plants—crops, medicinals, or spices, some used by other great cultures such as China and India. Also by that time the invention of printing allowed for publicizing the herbals. This led to an increasing interest in plant identification and to the search for new medical plants, which in turn led to the founding of chairs of botany at European medical schools, the first one in Padua, Italy, in 1533 (Mayr 1982).

The plants described in the herbals were not arranged in anything resembling today's classification with the exception of the latest herbals; for example, C. Bauhin's *Pinax*, published in 1623 (Bauhin 1623), contained not only a great number of different kinds of plants but their arrangement often reflects their kinships, close to our current perception. A great step in classification was made by Cesalpino with his publication *De plantis* (Cesalpino 1583), which influenced botany for the next two hundred years (Mayr 1982). Another invention contributed enormously to the advance of botany; this is the technique of pressing and drying plants, attributed in Mayr (1982) to Luca Ghini (1490–1556) as the basis of herbarium collections. Luca Ghini also established in 1543 the first university botanic garden in Pisa, Italy (Mayr 1982). Thus the concept of herbarium collections and botanic garden collections dates back to the sixteenth century.

Clerics were studying and cataloguing biological diversity as a way of celebrating and better understanding what they believed to be the master plan of their chosen deity. The missionary passion of the nineteenth-century British and French collectors was based on a similar view, and as a way of expressing the richness of their countries imperial view of the world. Medical people collected extensively in the pursuit of useful healing compounds, for example, Asa Gray, George Engelmann, and Sir John Richardson.

Linnaeus (1707–1778) laid the foundation for the binomial nomenclature (1753) that is currently practiced. However, although the generic concept goes back to Greek and Roman times, the modern genus concept originated with Tournefort (1694). The consistent use of concepts of genus and species enabled cataloguing of the diversity discovered as a result of many new voyages of exploration in the eighteenth century, for instance, those by the well-known Captain Cook. These voyages included a botanist as an important member of the expedition. Collections were made possible not only through the support of royalty but by the growing number of aristocrats or wealthy people who became interested in the pursuit of knowledge of biodiversity or who supported amateur botanists, for example, Sir Joseph Banks (1743–1820) in England, George Clifford (1685–1760) in Holland, the Jussieus (1686–1853) in France, Kunth (1788–1850) in Germany, and Mociño in Mexico (1700).

These collections became the basis for what we know now as the major museum collections. In the eighteenth and nineteenth centuries an increasing number of avid amateurs contributed enormously to these collections. Among these amateurs, some were able to finance their voyages and their own research because of their inherited personal wealth, for example, Boissier (1810–1885). Others were dedicated enough that they sought funds to pursue their interest in taxonomy by selling duplicates of their collections to museums, and often ended their lives in poverty, for example, Kotschy (1813–1866), who collected extensively (300,000 numbers, according to Rechinger 1960), especially in the Near East.

Irrespective of financial background, early explorations all over the globe were carried out by Europeans from Linnaeus's pupils and on into the European colonial period. Subsequently major centers of herbarium collections also became established in North America. These collections served as the basis for our knowledge of the flora, as well as for an understanding of the organization of the green world on which we depend so much. In most countries the major herbaria eventually became government supported from the nineteenth century on. An increasing decline in support is becoming critical for their survival, however.

The knowledge that has accumulated in the last two hundred years was made possible mainly through the efforts of committed botanists, often amateurs, and by and large without adequate support from society (e.g., government agencies). Today collecting is carried out by specialists for a particular purpose: by amateurs for recreation or self-improvement and by natural resource and conservation specialists in support of protection programs. Let us now review the purposes of collections.

Purposes of Collections

Collecting is carried out for completing the inventory of our knowledge and for acquiring information about traits and variation for utilization, preservation, and conservation.

Inventory of Species

The acquisition of knowledge of all taxa on Earth, particularly species, is achieved through species inventory. Since the type method is used for the application of species names, type collections are of primary importance and serve humanity as standards for plant names. [This is what taxonomists call the "Type method"; it is based on the second principle of the Botanical Code (Greuter et al. 1988:3): "The application of names of taxonomic groups is determined by means of nomenclatural types." A plant's name is the key to its literature (Davis and Heywood 1963) and to its uses in ancient days until the present.] Habitat destruction, especially in the tropics, has recently become a significant issue, and has exerted pressure particularly on taxonomy, which has been called on to inventory all species before many disappear (Campbell 1989; see also Mission 1 of Systematics Agenda 2000 [Anonymous 1994:1]). But as we shall see later, taxonomic methods using different approaches are necessary for setting up priorities for conservation of geographic areas, and the making of inventories alone is not enough. Preserved material is useful for determining undetected habitats of ecological significance. For instance, Catling, Catling, and McKay-Kuja (1992) were able to describe a disjunct area of prairie vegetation in southeastern Ontario that was destroyed a number of years ago.

Traits, Variation, Utilization, and Conservation

Knowledge of the different names of species alone is not enough. Different species vary considerably in morphology, ecological preferences, distribution, anatomy, cytology, chemistry, and in many different molecular traits. Some of these traits have proven highly useful for humankind, whereas others are believed to be potentially useful. The crop species sustain humanity, ornamentals beautify our cities and dwellings, medicinal plants were used since the dawn of civilization, lumber is used for dwellings, furniture, and art. The uses of course include fodder, fuel, environmental stabilization, and many others. The Tzeltal, for instance, a Mayan people from highland Chiapas, Mexico, live in an area of about 60 thousand km^2 and use many of the native plants, as well as recently introduced plants. Berlin et al. (1974) list 419 wild plants considered useful by the Tzeltals, including 60 now protected plants, and 96 cultivated plants. For instance, *Phytolacca rugosa* is

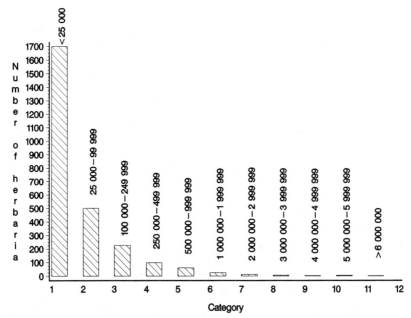

FIGURE 4.2   Herbaria of the world categorized by the size of their collection holdings.

collected for its edible leaves and fruits, and is also used as shampoo by Tenejapa women (Berlin et al. 1974).

In cultivated crops we are particularly interested in the infraspecific genetic diversity of the crop species and their wild relatives, as opposed to taxic diversity for inventory purposes. Further, species are increasingly being used today in the wild (taxic diversity) as environmental bioindicators.

Status of Collections

Herbaria

There are 2,639 herbaria with official status in the world (Holmgren et al. 1990:482). Of those, 1,700 hold less than 25,000 specimens each. In addition there are thousands of private herbaria of various sizes. The smallest public herbarium is the Seychelles with 500 specimens. The herbaria are summarized by size categories of their holdings (fig. 4.2).

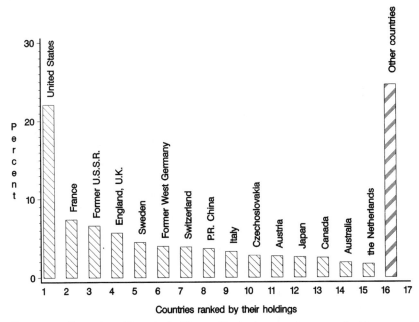

FIGURE 4.3   Countries with major holdings of herbarium specimens, ranked by percent of the world total.

Only two herbaria have a collection of 6 million specimens or more; these are the Musée National d'Histoire Naturelle, Paris, France, with 8,877,300 and the Royal Botanic Gardens, Kew, England, with 6 million. There are five herbaria containing between 5 and 6 million specimens— the Komarov Institute in St. Petersburg; the Swedish Museum of Natural History in Stockholm; The New York Botanical Garden, Bronx, New York; the Natural History Museum, London; and the Conservatoire et Jardin Botaniques, Genève.

A summary by country shows that the major holder of herbarium specimens is the United States (fig. 4.3), which holds 22%, whereas China, Japan, Canada, Australia, and a number of European countries hold between 2.5% and 8% each. All other countries together hold 25% (see fig. 4.3). The breakdown by continents and insular regions (fig. 4.4) shows that Europe is the major holder. Europe also leads in the number of public herbaria. The African continent holds the least number of specimens, but Australasia and the west Pacific islands, such as Borneo, Indonesia, and so on, have the least number of herbaria (fig. 4.5).

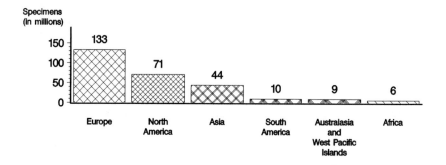

FIGURE 4.4   Herbarium collections of the world by continental and insular regions. *Source*: Data from Holmgren et al. 1990; map data from SAS/GRAPH, SAS™.

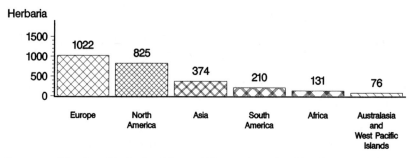

FIGURE 4.5   Number of herbaria of the world by continental and insular regions. *Source*: Data from Holmgren et al. 1990; map data from SAS/GRAPH, SAS™.

The European continent far exceeds (by 790,000) the other continents in the number of types combined that are contained in the herbaria (fig. 4.6). This is understandable given the early start and the continued interest in taxonomy in Europe, as mentioned in the historical perspective above.

Wilson (1985) estimates the total number of plant species at 440,000, and Grant (1991) provides a more conservative estimate with a breakdown, summarized here (fig. 4.7). Based on Grant's estimates there are 952 specimens per species on average. It should be remembered that specimens were often collected in duplicates, sometimes in batches of 100 for exchange (known as "centurias" or "exsiccatae"). Often one finds specimens collected from the same locality year after year by the same or many different collectors following one another's collecting

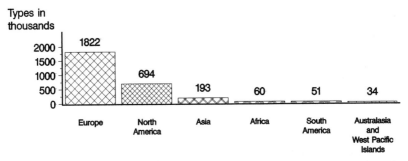

FIGURE 4.6   Type specimens held in herbaria of the world by continental and insular regions. *Source*: Data from Holmgren et al. 1990; map data from SAS/GRAPH, SAS™.

route. This average of 952 specimens per species tells us something, but we know that collecting was not done with evaluating variation or infraspecific diversity in mind.

The number of nomenclatural type specimens found in the herbaria of the world is 2,852,695 (Holmgren, Holmgren, and Barnett 1990). Based on Grant's (1991) conservative estimate of species (fig. 4.7) and given the number of types, there are about 10 type specimens for every species. This of course includes types of subspecies, varieties, and other infraspecific categories. It certainly includes the type of many synonyms. Thus on average, 1 in every 95 herbarium specimens is a type of some sort. A type includes not only what is known as the holo-type or lectotype, but often also many paratypes or syntypes as well as duplicates of those. Eleven countries contain more types than the gross average of 1 in every 95 specimens (fig. 4.8). The United States con-tains 1 type in every 90.41 and is close to the world average but con-tains by far the largest number of specimens (fig. 4.8, lower left). Austria holds 10.6% of all the types in a total of over 7 million speci-mens (fig. 4.8, near left axis); Senegal has 16.9% of types out of only 118,000 specimens (fig. 4.8, near left axis); and the United Kingdom holds many types, 57.9% of all the types known, but has 18.5 million specimens in many collections. These figures are not adjusted, but they tell us something about the number of type specimens in collec-tions of a country relative to their total collection holdings. The plot of number of specimens to the ratio of types per number of specimens (fig. 4.8) shows that about half the countries in the world hold at least 1 type of specimen in 1,000 or less specimens in their collections but that only a few countries have many types and large collections (fig.

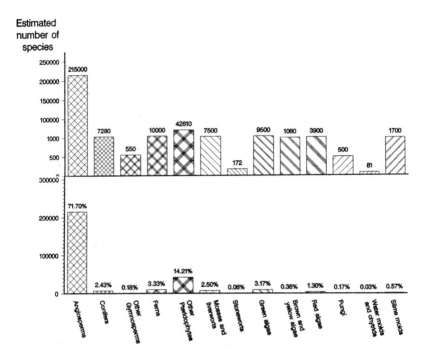

FIGURE 4.7   Species in different plant groups, estimated by Grant (1991) (*top*, number of species; *bottom*, percent).

4.8, lower left). Some of the major herbaria that also contain many types are listed in table 4.1.

On average, collections are very poor representations of plant variability from the statistical point of view. The tendency is to collect extremes and outliers in addition to typical specimens. Collecting is, as a rule, not done according to statistical sampling requirements (see below). Representation is critical for the utility of the collections, especially because of uneven collecting throughout the areal of every species.

Gene Banks (Germplasm Collections of Crops, Landraces, and Their Wild Relatives)

Approximately 3.5 million accessions, including duplicates, are maintained as seed in 121 countries, and regenerated. This includes the 18

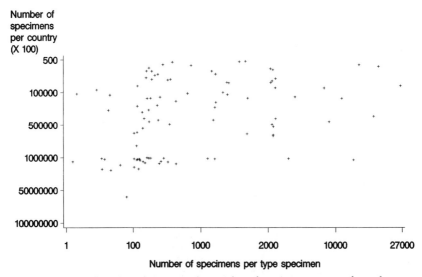

FIGURE 4.8   Plot of number of specimens by number of specimens per type for each country. Note, for example, that the two points closest to the vertical axis are two countries that have slightly more than 10 specimens per type, but one country has more than a million specimens while the other has 100,000.

research centers which together make up the Consultative Group on International Agriculture Research (CGIAR) (fig. 4.9), but only 11 of them maintain genetic resources. According to estimates, the CGIAR network holds up to 504,000 samples (M. Perry, pers. comm.), that is, 14.4% of the world's crop genetic material. The research centers distribute more than 100,000 samples to breeders worldwide annually. Seed preservation, whenever possible, is the most commonly used method of preservation, being both economical and practical. Among these seeds are many long-discarded landraces.

Most of the maintained accessions represent the seven most important crops ranging from 100,000 for sorghum to 250,000 for wheat (fig. 4.10). These seven crops are *Glycine max*, *Hordeum vulgare*, *Oryza sativa*, *Sorghum bicolor*, *Triticum aestivum*, *Triticum* unidentified species, and *Zea mays*, totaling 1.35 million accessions, which make up about 36% of all accessions maintained worldwide. A similar total number of accessions consists of taxa with accessions in the 10,000–100,000 range as shown in fig. 4.11. This group includes 53 items with taxa identified to species such as *Phaseolus vulgaris*, *Triticum durum*, *Pisum sativum*, and *Avena sativa*; others identified to genus only such as *Hordeum* spp., *Avena* spp.

TABLE 4.1

Major Herbaria with Both Many Collections and Many Type Specimens

Country	Place	Year	Specimen
Austria	Wien (W)	1807	3,750,000
Belgium	Meise (BR)	1870	2,040,000
Czechoslovakia	Praha (PRC)	1775	2,000,000
Denmark	Copenhagen (C)	1759	2,223,812
Finland	Helsinki (H)	1750	2,720,000
France	Paris (P)	1635	8,877,300
Germany	Berlin (B)	1815	2,500,000
	Jena (JE)	1895	3,000,000
Italy	Firenze (FI)	1842	3,600,000
Netherlands	Leiden (L)	1829	3,000,000
Russia	St. Petersburg (LE)	1823	5,770,000
Sweden	Lund (LD)	1770	2,400,000
	Stockholm (S)	1739	5,600,000
	Uppsala (UPS)	1785	2,500,000
Switzerland	Geneve (G)	1824	5,000,000
United Kingdom	Kew (K)	1841	6,000,000
	British Museum (BM)	1753	5,200,000
	Edinburgh (E)	1839	2,000,000
	Cambridge (GH)	1864	4,607,000
United States	Chicago (F)	1893	2,415,000
	New York (NY)	1891	5,300,000
	Saint Louis (MO)	1859	3,700,000

Source: Holmgren et al. (1990).

Hevea spp., and *Saccharum* spp.; and other taxa identified to family such as legumes. These 53 items make up almost 38% of all the germplasm accessions worldwide (fig. 4.11). The majority of the taxa (mostly at the species level) in the collections amount to one single or very few accessions (less than 500 in fig. 4.10).

Now, when we look at the distribution of the accessions maintained by different institutions by world sectors we see that Europe, including the Mediterranean region, maintains the highest number of accessions (fig. 4.12). The highest number of taxa found in gene banks is in North America (fig. 4.13), particularly in the United States. But North America has fewer gene banks than do most other continents (fig. 4.14). There is an almost equal number of gene banks in northern and central Europe as in the Mediterranean region (fig. 4.14). The high number of accessions in Asia contained in the many gene banks is the result of a large number of accessions of a few major crops, such as 81,500 accessions of *Oryza sativa* in the Philippines and 31,817 accessions of *Sorghum* spp. in India, of which some are specific to a particu-

FIGURE 4.9 The CGIAR global germplasm network. Abbreviations: CIAT—Centro Internacional de Agricultura Tropical; CIMMYT—Centro Internacional de Mejoramiento de Maiz y Trigo; CIP—Centro Internacional de la Papa; COFOR—Center for International Forestry Research (location not yet determined); IBPGR—International Board for Plant Genetic Resources; ICARD—International Center for Agricultural Research in the Dry Areas; ICRAF—International Council for Research in Agroforestry; ICRISAT—International Crops Research Institute for the Semi-Arid Tropics; IFPRI—International Food Policy Research Institute; IIMI—International Irrigation Management Institute; IITA—International Institute of Tropical Agriculture; ILCA—International Livestock Center for Africa; ILRAD—International Laboratory for Research on Animal Diseases; INIBAP—International Network for the Improvement of Banana and Plantain; IRRI—International Rice Research Institute; ISNAR—International Service for National Agricultural Research; WARDA—West Africa Rice Development Association. (Source 1992)

lar country such as the 30,000 collections of *Malus pumila* X *M. robusta* in China.

It is encouraging that there are relatively many gene banks in all continents. It is disappointing that so little effort in general is put into the research, preservation, and maintenance of crops, especially their wild relatives that do not fall into the category of major crops. This, however, does not mean that sufficient resources are allocated to the few major crops, especially their wild relatives. According to the International Board for Plant Genetic Resources (IBPGR; now International Plant Genetic Resources Institute, IPGRI) estimates that germplasm from wild species makes up less than 2% of the gene banks' holdings. What

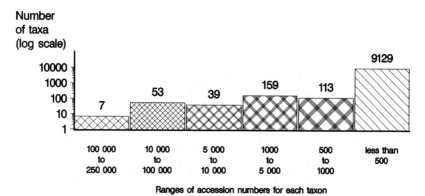

FIGURE 4.10   Ranked number of accessions by taxa in gene banks. See text for the seven taxa with the highest collections. *Source*: IBPGR database files.

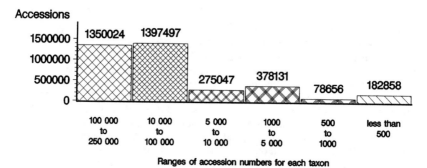

FIGURE 4.11   Ranked number of accessions by total number in gene banks. The total number includes duplicate accessions. Note that for the seven taxa with the highest collections (fig. 10) there are 1,350,000 accessions. *Source*: IBPGR database files.

is also disappointing is that 35–50% of the accessions within world collections are indiscriminate duplicates (Plucknett et al. 1987; Adham and van Sloten 1990). There may be significant collections for other more neglected crops, but the data are not available.

Botanic Gardens

Botanic gardens now also recognize a need to become involved in preservation in addition to their many other functions and responsibil-

Accessions

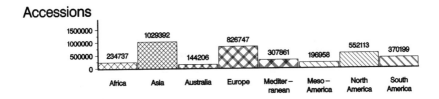

FIGURE 4.12 Summary of collections held in gene banks, including duplicates, by continent or region. *Source*: IBPGR database files.

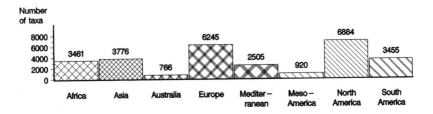

FIGURE 4.13 Summary of number of taxa in gene banks by continent or region. *Source*: IBPGR database files.

ities. There are approximately 1,300 botanic gardens in the world according to the recent International Directory of Botanical Gardens (Heywood, Heywood, and Jackson 1990), of which many recently opened in the tropics or subtropics where taxic diversity is highest. Botanical gardens have increasingly been engaged in *ex situ* conservation of threatened species. In this regard they play a similar role to zoological gardens.

A brief overview on botanical garden collections was provided by Heywood (1992). Based on a printout of the most recent raw data on botanic gardens accession numbers (courtesy of Dr. V. H. Heywood) the following summary is presented here. Europe has by far the highest number of botanical gardens (fig. 4.15). Countries in the north temperate regions have most of the botanic gardens, which is in sharp contrast to the richness of the floras, mainly in the tropics including South America (fig. 4.15). The number of accessions shows a similar picture. It must be remembered that there is much duplication as is the case in herbaria and in gene banks. Furthermore, a number of botanical gardens contain special collections of their region, such as the University of

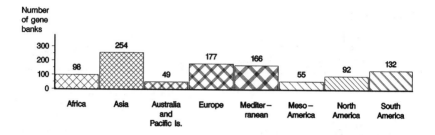

FIGURE 4.14 Summary of number of gene banks by continent or region. *Source*: IBPGR database files.

British Columbia Botanic Gardens, Canada, with about one-third of the native species of British Columbia that it maintains from a total of 14,000 taxa it exhibits. In Australia, although the number is relatively small compared to Europe (fig. 4.16), a recent estimate suggests that between one-third and one-half of the native flora of 20,000 species is maintained in that country's botanical gardens (Heywood 1992).

Census of botanic gardens is very difficult to conduct. Some gardens are sometimes not considered as such, and others do not maintain an inventory. For instance, the city of Chengdu, Sichuan Province, China, maintains a number of botanic gardens, one of which specializes in bamboos with hundreds of species cultivated in it. This garden is one of many for which data were unavailable in the raw data listing obtained from IUCN (International Union for the Conservation of Nature) and in which many gardens are listed without figures on number of taxa or accessions.

Regrettably there is an inverse proportion between the number of botanic gardens and patterns of plant diversity for a given area (see fig. 4.15), such as in South America where Brazil and Colombia, with a combined number of species of approximately 90,000, have only about 20 botanic gardens (Heywood 1990). This situation is rapidly changing, with the rate of new gardens being created increasing on almost a weekly basis in tropical countries (Heywood 1990). These estimates, however, are much too high and the rapid creation of botanic gardens in the tropics does not hold (P. H. Raven, pers. comm.). Thus these statistics are not very informative, but they are the best we have.

Botanic gardens have had varying activities with medicinal plants that go back a long time in history (see above). It is estimated that 25,000

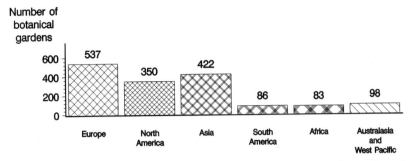

FIGURE 4.15 Number of botanical gardens in different continents. *Source*: Botanic Garden data courtesy of Plant Conservation Office, IUCN; map data from SAS/GRAPH™.

species have been used for medicinal purposes and that still about 80% of people, especially in developing countries, currently use them in different forms. The increasing demand for some of the medicinal plants is responsible for the drastic reduction of natural populations of some of these species. A case in point in Canada is the Ginseng (*Panax quinque-folium* L.), with not only a medicinal demand but a commercial one too. It is highly valued in the Orient as a panacea and is increasingly used in North America as a herbal medicine (Argus and White 1984).

Thus botanic gardens increasingly undertake the preserving of rare and endangered species. Unfortunately, the majority of the species are maintained by a single accession. Thus, because of their lack of infra-specific diversity, botanic gardens do not come close to the goals of genetic resources collections (gene banks).

Sampling and Measures of Diversity

Sampling

For collections to be useful for research and conservation they must be truly representative of the genetic diversity. To achieve this they ought to be based on sound sampling practices. Sampling is carried out in order to obtain information about the statistical universe or population of interest in such a way that the inferences drawn from the samples are accurate enough to reflect the situation from which we have drawn the

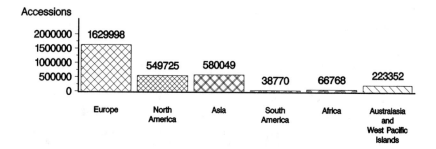

FIGURE 4.16   Summary of accessions/taxa reported by botanical gardens in each conti-nent. *Source*: Botanic Garden data courtesy of Plant Conservation Office, IUCN; map data from SAS/GRAPH™.

samples. Sampling is needed because of nonstatistical considerations, such as cost, time, available resources, and prior information on hand. Essentially, sampling consists of obtaining information from a portion of a larger group, or "universe" (Slonim 1966). The method used for sampling, or a combination of methods used, will be determined by the problem one wishes to solve. Sampling enables one to cut the cost of gathering information from the complete "statistical universe." By information I mean the data as well as the statistics about the problem area. By choosing a particular method of sampling and scoring of the traits or stimuli, and subsequently computing statistical estimates, we aim to portray, as closely as possible, the real statistics of the problem area. In taxonomy this is genetic variability, as in the study of crops and their wild relatives, and taxic variation for most groups of plants.

Various classical approaches to sampling are available. They fall into the following categories: random sampling, stratified sampling, systematic sampling, cluster and multistage sampling, and combinations of approaches.

In simple random sampling every unit should have the same probability of being selected in the sample. This requirement is never met in herbarium collections and is rarely done in plant surveys.

Stratified sampling requires some knowledge about the strata, and then a random sample is taken of each. This is fine when one already has knowledge about the total inventory of the kind of species and genera (the strata) in a particular area. Variations of this sampling method are (1) proportional stratified sampling and (2) equal size stratified sampling. There are further variants within each of these. In each variation one would again pick a random sample. A determining factor is the time and cost given to achieve accuracy of the estimates. However, in biology we seek precision as much as possible in order to achieve accuracy. Unfortunately precision requirements are negatively correlated with budgetary constraints.

In systematic sampling the units are sampled according to previous knowledge of the kind of units to be sampled, but every rth individual is selected among, say, all the individuals of a particular species. This is never done in systematic studies and cannot be accomplished given time and budgetary constraints. One exception is the careful and akin to systematic sampling carried out on *Sedum* of the Trans-Mexican volcanic belt by Clausen (1959).

In cluster sampling and multistage sampling the units can be divided into clusters by some criteria, such as area of distribution or habitat. Then a sample is drawn from the clusters. In the case of multistage sampling, units are chosen in proportion to some criterion, for example, flowering time variation within a species. This requires information, some of which may be obtained from specimens in collections.

Sizes of the Samples

There are some widely held misconceptions about sample size requirements, such as that "to be accurate a sample should include at least 10% of the population or the universe" or "the more individuals we study the more accurate our investigation will be." In other words, the popular misconception is that the size of the population is what determines the number of individuals to be sampled from that population. As it turns out, though, the emphasis ought to be placed on the number of individuals in a sample and not on the population or the universe (fig. 4.17). For instance, if the population size is 100, the number of individuals required to be sampled is 79 for a permissible error of 0.05 and 99 for a permissible error of 0.01. The number decreases relative to population size until it reaches a plateau of around 320 individuals irrespective of population size for a 0.05 error and around 9,500 for a population above 500,000 individuals as shown in the curves (fig. 4.17).

The size of samples is often neglected or it is a contentious issue. Most biosystematic research is indifferent to this aspect, and in some cases available resources are a limiting factor. To distinguish between hypotheses, say, whether two groups are different at the species level or

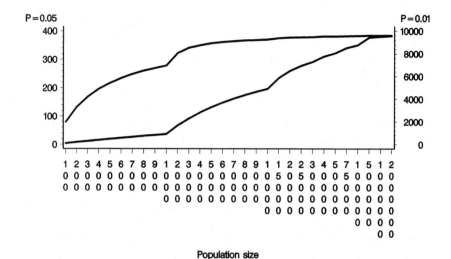

Population size

FIGURE 4.17 Size of sample required to represent a given population size, given random sampling with 95% confidence level, for 0.05 and 0.01 proportion permissible error. Note that for a population of more than 3,000 the sample size required is close to 350 for *P* = 0.005, levels to 9,700 for a population of half a million and above for *P* = 0.01. *Source:* McCall (1982) Table H.7.

below such as hybrid status, requires large samples. The sample size can be determined for each approach by statistical requirements; see, for instance, Milligan (1992) regarding organelle DNA. In a recent debate regarding the survey of molecular genetic diversity of human-kind, a compromise was reached to survey blood samples of 25 individuals per population to allow surveying 400 populations out of the 5,000–7,000 believed to exist (Roberts 1992). Sample size is also related to sampling criteria, which is relevant to molecular studies (Kohn 1992).

Measures of Diversity

There are different approaches to measuring diversity. The distinction between different purposes ought to be emphasized here. One might distinguish between taxic diversity and genetic diversity. The latter is mostly at the population and infraspecific level, suitable for evaluation of germplasm of cultivated crops and their wild relatives.

Taxic diversity. These measures are aimed at conservation of genetic resources of geographic areas. Most are based on straight inventories, species richness, and a combination of richness and abundance. For tropical areas, where inventories have not yet been completed, Campbell (1989) used the number of collections per 100 km^2 per year for a given area or country to assess the number of years required to reach the level of 100 specimens/100 km^2, considered to be adequate for a given area, based on the rate of collecting in 1974–1981 taken from *Index Herbariorum*. Taxic diversity obviously cannot be studied from areas lacking adequate collections. According to Campbell's figures, a number of tropical countries appeared to be sufficiently collected for the purpose of determining conservation areas, such as Malaysia, Singapore, Taiwan, and Costa Rica. But at the other extreme are countries that appeared to be far off from reaching that stage, such as The Guianas, Papua New Guinea, Burma, and Bolivia.

Margules, Nicholls, and Pressey (1988) developed an algorithm to help determine maximum biological diversity, in terms of species presence, in a particular or local ecosystem to be used as criteria for deciding on the extent of sets of sites in a geographical area to set aside for conserving the specific ecosystem. An iterative method was developed earlier by Kirkpatrick (1983). Coddington et al. (1991) tried a number of approaches at estimating species diversity in tropical Bolivia and tested them. They found that collector, time of day, and method of collecting have significant effects on abundance.

Recently measures of phylogenetic (taxic) diversity as a basis for setting priorities for conservation of geographic areas, requiring fully resolved cladograms, have been proposed. May (1990), Vane-Wright, Humphries, and Williams (1991), and Williams, Humphries, and Vane-Wright (1991) developed indexes based on cladograms of taxa to be assessed for conservation. Each terminal taxon is ranked according to taxonomic statements of belongingness to the different clades on a particular cladogram which is then normalized. Vane-Wright, Humphries, and Williams (1991) went one step further in using their rankings for their priority area analysis using the taxic weights from their previous process of ranking. This is done by scoring the representation of the terminal taxa in different a priori defined geographical areas on the cladogram, summing up the scores for each taxon. The region with the highest score is then adjusted for sympatry to determine the most desirable region for conservation.

Another phylogenetic approach, but one not requiring the full resolution of cladograms, aims to minimize diversity lost and maximize diversity preserved. In this approach the diversity of a group must be evaluated relative to sister groups, not in absolute terms. Nixon and Wheeler (1992) presented two such measures that depend on the estimated number of terminal taxa for each clade. Their measure yields a ranking of taxa. The ranking provides a guideline but does not determine the actual areas nor the specific species to conserve.

Genetic diversity. For cultivated plants four functional classes were introduced by the International Biological Program (IBP) in 1966: landraces, advanced cultivars, wild relatives of domesticated plants, and wild species used by man.

A number of countries have made an effort to preserve their landraces as sources of genetic materials for plant improvement. In many cases the sampling was inadequate as it was far from being representative of the genetic diversity. The gene pool of landraces has for most common crops been drastically reduced. Those that persist occur in isolated, remote, and inaccessible areas (e.g., Hammer et al. 1981; Perrino, Hammer, and Lehmann 1982). To conserve landraces entails the conservation of the conditions of traditional agricultural ecosystems.

Advanced cultivars, that is, those bred with modern methods, have a narrow genetic base. Many older cultivars are replaced by varieties of "superior quality" without being preserved. As a result, some genes are lost forever.

Wild relatives of cultivated plants harbor a wealth of genetic material for crop improvement and development. Genetic analysis in wild populations has shown that sampling strategies depend on the breeding system, such that the size of a representative sample increases from apomicts through cross-fertilizers to self-fertilizers (Whyte 1958). Multipurpose sampling is difficult, although some basic approaches were suggested (Allard 1970; Bennett 1970; Qualset 1975). Bennett emphasized random sampling and adding "rare" types to the sample. Yonezawa (1985) recommended ten plants per site and maximizing the number of sites, akin to cluster sampling. Maximizing the number of sites with an undefined minimum number of plants was found to be a practical compromise of the scientific targets taking the costs into consideration by Baum et al. (1975, 1984). Qualset suggested a minimum of five hundred plants to capture the variation in each population.

Wild species used by humans include many forest trees. Forest species are used for food, such as fruits, oils, beverages, and spices. Numerous native stocks of apples, pears, plums, apricots, almonds, and grapes have disappeared in their centers of diversity in the Near East

TABLE 4.2
Germplasm Collections, Live and Cryogenic

1. Apples; 76 major apple germplasm programs; North America, Europe, Japan, and China
2. Apricots; Soviet collections and Europe
3. Cherries; Europe
4. Peaches
5. Plums; 91 collections for 35 countries
6. Grapes; 14,000 cultivars
7. Blackberries and raspberries
8. Blueberries and cranberries
9. Currants and gooseberries; many collections
10. Strawberries
11. Amelanchier
12. Pawpaw; no collection for the specific purpose of germplasm maintenance
13. Kiwi fruit; China has many collections, others in New Zealand and the United States
14. Pears; 67 collections in 32 countries, also the USSR, China, Japan, and the United States
15. Almonds; 30 world collections in 21 countries
16. Chestnuts; many collections but few approach the standards for germplasm maintenance; China, France, Italy, Japan, Korea, Turkey, and the United States
17. Hazelnuts; 400 distinct cultivars, only half in collections; 22 countries 39 collections
18. Pecans and hickories; more than 1,000 cultivars; the United States, Canada, Mexico, and a few other countries maintain collections
19. Walnuts; the United States, Italy, and France have major germplasm maintenance collections

Note: Unfortunately there are no summaries. In many of the crops listed in the table, many wild relatives are poorly collected.

Source: Moore and Ballington (1990). E. Bettencourt and J. Konopa authored the *Directory of Germplasm Collections*, published by IBPGR.

and Central Asia. A number of preservation programs of cultivars, including early stocks, are in place (see above, and table 4.2). For a number of species it is too late for in situ conservation. For many other trees, especially the tropical ones, conservation is possible. For example, most of the native germplasm of the cashew nut in its native tropical American region has never been utilized, since approximately 90% of its commercial production is in East Africa and India and stems from a narrow genetic base (Kumaran, Nayar, and Murthy 1977).

Adequacy of Collections

In view of the brief general discussion of sampling requirements for different purposes presented above, it is clear that the three main collecting thrusts, namely, gene banks, botanical gardens, and herbaria, although complementary, are unable to fulfill the desirable statistical requirements as depositories of material for future research needs. Gene banks of cultivated crops and their wild relatives preserve some genetic diversity, including in some cases landraces that are now practically extinct in some major crops. Botanical gardens preserve some genetic diversity of selected native plants. Herbaria preserve expressions of morphological variations within and between species and as reflecting geographic distribution, ecological isolation and variation, dispersal mechanisms, taxonomic age, and so on. Herbaria also serve to assist in the process of determining areas to set aside for conservation of diversity, with ecological integrity, size, representation, and other special features combining to identify significant natural environment sites. They bear the testimony of the present state of our knowledge of species (taxic) diversity and our knowledge of the different floras. Herbarium collections inadequately represent variation below the species level. There is indeed no substitute for the biodiversity present in natural habitats as a source of germplasm for the future. In this sense the distinction between preservation and conservation is paramount.

We have seen that the required reasonable sample size, when the choice is simple random sampling, is 300–400 individuals per population to achieve reasonable precision in the estimation of basic variation statistics. To describe a species it is desirable to have representatives from different locations across some kind of a grid of the areal of the particular species. For a simple diagnosis of a species, this amount of sampling was never required. One attempt in this direction was carried out by Clausen (1959) in his work on *Sedum*, mentioned earlier. In the study of biodiversity, especially of cultivated crops and their wild relatives, for the purpose of evaluation and utilization of genetic material, it is not only desirable but necessary to apply sampling methodology as required.

Most herbarium collections are barely adequate for detailed studies of biodiversity at more than a geographically limited level. This is the result of many factors, first and foremost being the lack of financial support.

Although herbarium collections are very useful for taxonomic studies, their utility is presently limited for some newer approaches dependent on fresh genetic material. Even though DNA can be extracted from herbarium specimens, it is useful only in a limited way. For the study of biodiversity one must resort to collecting live material in the field and over a representative area, using adequate sampling methods. One of the main causes of unsatisfactory classification, listed by Davis and

Heywood (1963), is insufficient material, mainly because of inadequate collecting or sampling.

I feel it is unreasonable to expect gene banks, botanical gardens, and herbaria to fulfill the statistical need of sampling requirements. Clear goals and priorities have to be set. Wilson (1988) was one of the initiators of an opinion held by a growing number of taxonomists that it is important to characterize biodiversity in terms of species inventories, including fossils, and their phylogenetic history, and that the task might not be accomplished in view of the global, human-mediated, and accelerating rate of extinction. In this regard Novacek and Wheeler (1992) stress the importance of taxic versus character sampling in connection with phylogenetic history.

The requirements for sampling at the DNA level are equally important. For example, the excellent work of Rieseberg, Choi, and Ham (1991) shows that faulty phylogenetic hypotheses based on chloroplast DNA data can be avoided if data from nuclear gene sequences are analyzed and after one devises comprehensive sampling strategies. Wheeler (1992) makes the point that drawing on characters from a single taxon to represent a large and diverse group could be viewed as a form of artificial extinction. Most molecular studies, often by necessity as a result of limited resources, are based on very limited sampling. In a simulation study Wheeler has shown that the accuracy of a cladogram is most influenced by the number of taxa, more so than by the number of nucleotide sites. He therefore argued that accurate results can be ensured by inclusion of the highest number of taxa in a particular group. Emphasis is here again put on taxic diversity as opposed to diversity within species, which is especially important in studies of plants of economic importance.

In sum, preservation of live material in the 1,023 gene banks and in the 1,300 botanic gardens, and the collections in 2,638 herbaria, despite their statistical inadequacy, are indispensable for continuous scientific activities in the improvement of the conditions of humankind.

These scientific activities will eventually influence society to introduce stringent conservation measures of the planet's biodiversity. Effective conservation of habitats will reduce the value of herbaria, gene banks, and botanic gardens for preservation, leaving their requirement strictly for research and education. Although there is no substitute for conserving native habitats and ecosystems, research in the phenomenon of past habitat destruction and species extinction will provide the tools for selective habitat conservation. Until such global measures of conservation are in place, a program of systematic collecting that provides a statistically valid base is needed for germplasm collections, botanical gardens, and herbaria. Increasing public awareness of the importance of germplasm for the food industry, in ethnobotany, and in ecotourism, and so on, will also raise the sociocultural value of collections. The future of collections will depend on partnerships to be developed between public and private sources.

Acknowledgments

The following colleagues provided useful comments for improving this manuscript: Dr. John Arnason, Department of Biology, University of Ottawa, Ontario; Dr. Peter H. Raven, director, Missouri Botanical Garden, St. Louis, Missouri; Daniel Brunton, Consulting Services, Ottawa, Ontario; and Dr. M. Perry, documentation coordinator, IBPGR, Rome.

I am most grateful to Professor Vernon Heywood, director of the Botanic Gardens Conservation International, Richmond, Surrey, for providing a listing of botanical gardens with numbers of holdings extracted from the BGCI database; and to Dr. M. Perry, documentation coordinator of IBPGR, Rome, for providing the entire IBPGR database.

Graphs (figs. 4.2–4.8 and 4.10–4.17) were accomplished using SAS/GRAPH, and SAS programs and map database, SAS®.

Literature Cited

Adham, Y. J. and D. H. van Sloten. 1990. The case for a wheat genetic resources network. In J. P. Srivastava and A. B. Damania, eds., *Wheat Genetic Resources: Meeting Diverse Needs*, pp. 139–144. London: Wiley.

Allard, R. W. 1970. Population structure and sampling methods. In O. H. Frankel and E. Bennett, eds., *Genetic Resources in Plants: Their Exploration and Conservation*, pp. 97–108. Oxford: Blackwell.

Anonymous. 1994. *Systematics Agenda 2000: Charting the Biosphere. Technical Report.* New York: Systematics Agenda 2000.

Argus, G. W. and D. J. White. 1984. *Panax quinquefolium* L. In G. W. Argus and C. J. Keddy, eds., *Atlas of the Rare Vascular Plants of Ontario*, part 3. Ottawa: Botany Division, National Museum of Natural Sciences.

Bauhin, C. 1623. Pinax Theatri Botanici . . . sive Index in Theophrasti, Dioscoridis, Plinii ei Botanicorum . . . Basel.

Baum, B. R., T. Rajhathy, J. W. Martens, and H. Thomas. 1975. *Wild Oat Gene Pool: A Collection Maintained by the Canada Department of Agriculture. Canada Avena (CAV).* 2d ed. Ottawa: Agriculture Canada.

Baum, B. R., R. von Bothmer, N. Jacobsen, G. Fedak, I. Craig, and L. G. Bailey. 1984. *Barley Gene Pool: A Collection Maintained by Agriculture Canada and by the Danish and Swedish Agricultural Universities (Nordic Gene Bank). Canadian-Scandinavian Hordeum Collection (CHC).* Ottawa: Agriculture Canada.

Bennett, E. 1970. Tactics of plant exploration. In O. H. Frankel and E. Bennett, eds., *Genetic Resources in Plants: Their Exploration and Conservation*, pp. 157–180. Oxford: Blackwell.

Berlin, B., D. E. Breedlove, and P. H. Raven. 1974. *Principles of Tzeltal Plant Classification: An Introduction to the Botanical Ethnography of a Mayan-speaking People of Highland Chiapas*. New York: Academic Press.

Campbell, D. G. 1989. The importance of floristic inventory in the tropics. In D. G. Campbell and H. D. Hammond, eds., *Floristic Inventory of Tropical Countries: The Status of Plant Systematics, Collections, and Vegetation, plus Recommendations for the Future*, pp. 5–30. New York: New York Botanical Garden.

Campbell, D. G. and H. D. Hammond, eds. 1989. *Floristic Inventory of Tropical Countries: The Status of Plant Systematics, Collections, and Vegetation, plus Recommendations for the Future*. New York: The New York Botanical Garden.

Catling, P. M., V. R. Catling, and S. M. McKay-Kuja. 1992. The extent, floristic composition, and maintenance of the Rice Lake Plains, Ontario, based on historical records. *Can. Field-Nat.* 106:73–86.

Cesalpino, A. 1583. *De Plantis Libri XVI. Ad serenissimum Franciscum Medicem, Magnum Etruriae Ducem*. Florence.

Cherfas, J. 1991. Disappearing mushrooms: Another mass extinction? *Science* 254:1458.

Clausen, R. T. 1959. *Sedum of the Trans-Mexican Volcanic Belt: An Exposition of Taxonomic Methods*. Ithaca, N.Y.: Comstock.

Coddington, J. A., C. E. Griswold, D. S. Davila, E. Peñaranda, and S. F. Larcher. 1991. Designing and testing sampling protocols to estimate biodiversity in tropical ecosystems. In E. C. Dudley, ed., *The Unity of Evolutionary Biology*. [*Proc. of the Fourth International Congress of Systematic and Evolutionary Biology. June–July 1990, College Park, Maryland.*] 1:44–60. Portland, Oreg.: Dioscorides Press.

Crepet, W. L., E. M. Friis, and K. C. Nixon. 1991. Fossil evidence for the evolution of biotic pollination. *Philos. Trans. Roy. Soc. London B. Biol. Sci.* 333:187–196.

Darlington, C. D. and E. K. Janaki Ammal. 1945. *Chromosome Atlas of Cultivated Plants*. London: Allen and Unwin.

Davis, P. H. and V. H. Heywood. 1963. *Principles of Angiosperm Taxonomy*. Princeton, N.J.: Van Nostrand.

Frankel, O. H. and E. Bennett, eds. 1970. *Genetic Resources in Plants: Their Exploration and Conservation*. Oxford: Blackwell.

Frankel, O. H. and M. E. Soulé. 1981. *Conservation and Evolution*. Cambridge: Cambridge University Press.

Grant, V. 1991. *The Evolutionary Process: A Critical Study of Evolutionary Theory*. 2d ed. New York: Columbia University Press.

Greuter, W., H. M. Burdet, W. G. Chaloner, V. Demoulin, R. Grolle, D. L. Hawksworth, D. H. Nicolson, P. C. Silva, F. A. Stafleu, E. G. Voss, and J. McNeill. 1988. *International Code of Botanical Nomenclature*. Königstein, Germany: Koeltz Scientific Books.

Hammer, K., M. Górski, P. Hanelt, F. Kühn, W. Kulpa, and J. Schultze-Motel. 1981. Variability of wheat landraces from Czechoslovakia and Poland. *Kulturpflanze* 29:91–97.

Harlan, J. R. 1961. Geographic origins of plants useful to agriculture. In R. E. Hodgson, ed., *Germplasm Resources*, pp. 3–19. Publ. 66. Washington, D.C.: Amer. Assoc. Adv. Sci.

Hey, J. 1992. Using phylogenetic trees to study speciation and extinction. *Evolution* 46:627–640.

Heywood, C. A., V. H. Heywood, and P. W. Jackson, comps. 1990. *International Directory of Botanical Gardens*. 5th ed. Königstein, Germany: Koeltz Scientific Books.

Heywood, V. H. 1990. Botanic gardens and the conservation of plant resources. *Impact of Science on Society* 158: 121–132.

——. 1992. Conservation of germplasm of wild plant species. In O. T. Sandlund, K. Hindar, and A. D. H. Brown, eds., *Conservation of Biodiversity for Sustainable Development*, pp. 189–202. Oslo: Scandinavian University Press.

Holmgren, P. K., N. H. Holmgren, and L. C. Barnett. 1990. *Index Herbariorum*. Part 1: *The Herbaria of the World*. 8th ed. New York: Intern. Asso. Pl. Taxonomy.

IBP. 1966. Plant gene pools. *IBP News* 5:48–51.

Kirkpatrick, J. B. 1983. An iterative method for establishing priorities for the selection of nature reserves: An example from Tasmania. *Biol. Conserv.* 25:127–134.

Kohn, L. M. 1992. Developing new characters for fungal systematics: An experimental approach for determining the rank of resolution. *Mycologia* 84:139–153.

Kumaran, P. M., N. M. Nayar, and K. N. Murthy. 1977. A study of the variation in cashew (*Anacardium orientale* L.: Anacardiaceae) germplasm, pp. 16–19. In *3rd Inter. Congress Soc. Adv. Breeding Researches in Asia and Oceania (SABRAO)*.

Linnaeus, C. 1753. *Species Plantarum*. Stockholm.

Margules, C. R., A. O. Nicholls, and R. L. Pressey. 1988. Selecting networks of reserves to maximize biological diversity. *Biol. Conserv.* 43:63–76.

May, R. M. 1990. Taxonomy as destiny. *Nature* (London) 347:129–130.

Mayr, E. 1982. *The Growth of Biological Thought*. Cambridge, Mass.: Harvard University Press.

McCall, C. H., Jr. 1982. *Sampling and Statistics Handbook for Research*. Washington, D.C.: National Education Assoc.

McNeely, J. A., K. R. Miller, W. V. Reid, R. A. Mittermeier, and T. B. Werner. 1990. *Conserving the World's Biological Diversity*. Gland, Switzerland: Intern. Union Conserv. Nature.

Milligan, B. G. 1992. Is organelle DNA strictly maternally inherited? Power analysis of a binomial distribution. *Amer. J. Bot.* 79:1325–1328.

Moore, J. N. and J. R. Ballington, Jr., eds. 1990. Genetic resources of temperate fruit and nut crops. *Acta Horticult.* 290 (in 2 parts). Wageningen, The Netherlands: Intern. Soc. Hort. Sci.

Nixon, K. C. and Q. D. Wheeler. 1992. Measures of phylogenetic diversity. In M.

J. Novacek and Q. D. Wheeler, eds., *Extinction and Phylogeny*, pp. 216–234. New York: Columbia University Press.

Novacek, M. J. and Q. W. Wheeler. 1992. Introduction. In M. J. Novacek and Q. D. Wheeler, eds., *Extinction and Phylogeny*, pp. 1–16. New York: Columbia University Press.

Perrino, P., K. Hammer, and C. O. Lehmann. 1982. Collection of landraces of cultivated plants in South Italy, 1981. *Kulturpfanze* 30:181–190.

Plucknett, D. L., N.J.H. Smith, J. T. Williams, and N. M. Anishetty. 1987. *Gene Banks and the World Food*. Princeton, N.J.: Princeton University Press.

Qualset, C. O. 1975. Sampling germplasm in a center of diversity: An example of disease resistance in Ethiopian barley. In O. H. Frankel and J. G. Hawkes, eds., *Crop Genetic Resources for Today and Tomorrow*, pp. 81–96. Cambridge: Cambridge University Press.

Raup, D. M. 1981. Extinction: Bad genes or bad luck? *Acta Geol. Hisp.* 16:25–33.

Raup, D. M. and J. J. Sepauski, Jr. 1984. Periodicity of extinctions in the geological past. *Proc. Nat. Acad. Sci., USA* 81:801–805.

Rechinger, K. H. 1960. Theodor Kotschy, ein pionier der Botanischen orientforschung. *Taxon* 9:33–35.

Rieseberg, L. H., H. C. Choi, and D. Ham. 1991. Differential cytoplasmic versus nuclear introgression in *Helianthus*. *Jour. Heredity* 82:489–493.

Roberts, L. 1992. How to sample the world's genetic diversity. *Science* 257:1204–1205.

Simberloff, D. S. 1986. Are we on the verge of a mass extinction in the tropical rain forest? In D. K. Elliott, ed., *Dynamics of Extinction*, pp. 165–180. New York: Wiley.

Slonim, M. J. 1966. *Sampling: A Quick, Reliable Guide to Practical Statistics*. New York: Simon and Schuster.

Tournefort, J. P. de. 1694. *Élémens de Botanique ou Méthode pour Connoître les Plantes*. Paris.

Vane-Wright, R. I., C. J. Humphries, and P. H. Williams. 1991. What to protect? Systematics and the agony of choice. *Biol. Conserv.* 55:235–254.

Wheeler, W. C. 1992. Extinction, sampling, and molecular phylogenetics. In M. J. Novacek and Q. D. Wheeler, eds., *Extinction and Phylogeny*, pp. 205–215. New York: Columbia University Press.

Whyte, R. O. 1958. *Plant Exploration, Collection, and Introduction*. Rome: Food and Agriculture Organization of the United Nations.

Williams, P. H., C. J. Humphries, and R. I. Vane-Wright. 1991. Measuring biodiversity: Taxonomic relatedness for conservation priorities. *Austral. Syst. Bot.* 4:665–679.

Wilson, E. O. 1985. The biological diversity crisis. *Bioscience* 35:700–706.

———. 1988. The current state of biological diversity. In E. O. Wilson and F. M. Peters, eds., *Biodiversity*. pp. 3–18. Washington D.C.: National Academy Press.

Yonezawa, K. 1985. A definition of the optimal allocation of effort in conservation of plant genetic resources with application to sample size determination for field collection. *Euphytica* 34:345–354.

5

Collecting Methodologies for Plant Samples for Pharmaceutical Research

JAMES S. MILLER

Numerous authors have documented the importance of plants as past and potential sources of therapeutic compounds. However, most efforts in drug discovery this century have focused on rational design synthesis and the production of synthetic medicines. The technology necessary for rapid, efficient evaluation of large numbers of biological samples has only recently become available, and this has led to a resurgence of interest in evaluating plants. Natural products drug-discovery programs have two major decision-making points: (1) the rationale and method by which samples will be collected and preserved for future evaluation; and (2) the method by which evaluation will take place. This paper examines possible strategies for sample acquisition and preservation. Given our poor state of knowledge of tropical biodiversity and the complex ecological interactions that characterize tropical ecosystems, the tropics may be expected to yield many interesting therapeutically useful compounds in the future. There is probably no sin-

gle optimal method for selecting species for evaluation, and the most appropriate methodology may depend on the goals of the research laboratory in which they will be evaluated. Determination of the optimal method for preserving the chemical integrity of samples would be extremely difficult but the available data support certain trends. Rapid drying, without exposure to extreme temperatures, is proposed to be the most efficient means of preserving samples for later evaluation.

Throughout the history of humankind, plants have been the major source of medicines. Natural products, including not only plants but microorganisms and animals, are very important to our pharmacopoeia. In a broad sense they have been estimated to be the source of more than 50% of clinically used drugs, with half of these from higher plants (Balandrin, Kinghorn, and Farnsworth 1993). The twenty top-selling drugs in 1988 resulted in 6 billion dollars in U.S. sales and, in one way or another, the discovery of all twenty is the result of natural products research (Miller and Brewer 1992).

Importance of Plant-Derived Medicines

Plants can be important sources of medicines in a number of ways. In addition to the use of plant-derived, purified chemicals that are used therapeutically in Western medicine, it has been estimated that about 80% of the population of the developing world relies on traditional medicine for their primary health care (Farnsworth and Soejarto 1985; Farnsworth 1990). For the 78% of the world's population that live in the developing world, that are projected to increase to 81% in 2010 and 84% in 2025 (Population Reference Bureau 1993), traditional medicine will necessarily continue to play an important role in primary health care.

In the past, plants have yielded many important drugs including, for example, vincristine and vinblastine (Noble, Beer, and Cutts 1958; Svoboda 1961) used to treat childhood leukemia and Hodgkin's disease, respectively. A more recent example is the taxane diterpenoid taxol (Kingston 1993), which, although its structure was first published in 1971 (Wani et al. 1971), was not approved for the treatment of refractory ovarian cancer until December 1992 (Balandrin, Kinghorn, and Farnsworth 1993) because of numerous problems, particularly with formulation, administration, and supply.

The Future Potential of Plants as Medicines

The past and present ability of plants to yield new medicines is well documented, and their future potential should not be doubted. The number of potential therapeutic areas in which each plant could be tested for efficacy is essentially limitless. For example, there are 4,200 phenotypically recognizable genetic diseases in humans (Webster et al. 1989). Likewise, cancer is more than 200 distinct diseases, each with a range of variability, and there are a myriad of known infectious diseases. Thus there is an almost infinite number of maladies against which the 250,000 known species of plants, with perhaps 10,000 still to be described (Gentry 1993), could yield new drugs. Plants have only begun to be evaluated and are expected to be a valuable resource in the future.

It has been nearly fifty years since Meyer (1946:346) wrote, "Where once it seemed that malaria was to be man's scourge for all of time, the hope is now dawning that someday malaria, like yellow fever, will have its ravages prevented." That there are more than 270 million new cases of malaria per year, resulting in between 2 and 3 million deaths (Klayman 1993), is illustrative of the point that diseases are seldom cured, and it is often necessary to react to resistance with new drugs. Thus research on artemisinin, derived from *Artemisia annua* L., may provide an alternative to the currently used antimalarials for strains of *Plasmodium* resistant to the drugs based on the structure of quinine.

Recent discoveries of the Natural Products Branch, Developmental Therapeutics Program, Division of Cancer Treatment of the United States National Cancer Institute (Cragg et al. 1995) are a further indication that plants will continue to yield medicinally useful compounds. *Ancistrocladus korupensis* D. Thomas and Gereau (Ancistrocladaceae), a recently described species from southwestern Cameroon (Thomas and Gereau 1993), has yielded the dimeric alkaloid michellamine B, which has shown promise as an anti-AIDS agent (Manfredi et al. 1991). Calanolide A, a coumarin isolated from *Calophyllum lanigerum* Miq. var. *austrocariaceum* (T. C. Whitmore) P. F. Stevens (Clusiaceae), is another recent discovery with anti-AIDS potential (Kashman et al. 1992). Additional promising compounds have also been discovered from *Omalanthus nutans* (Forster f.) Guill. (Euphorbiaceae), source of the phorbol ester prostratin (Gustafson et al. 1992), and a species of *Conospermum* Sm. (Proteaceae), which has yielded the trimeric naphthoquinone derivative, conocurvone (Decosterd et al. 1993).

Sampling the Plant World

The successful development of new medicines depends on the continued discovery of novel bioactive compounds, not merely extracts with bioactivity. The life of a drug consists of two phases, research and commercial (Miller and Harder 1994). In 1987 the research phase was estimated to cost 231 million dollars (Vagelos 1991) and was perhaps 300 million dollars in 1993 (J. Duke, pers. comm.). For this reason it is particularly important to maximize the efficiency of every stage of the drug-discovery process.

Most drugs on the market were discovered by large-scale screening projects (Bugg, Carson, and Montgomery 1993). The screening process has two major decision points: how samples will be acquired for screening, and how they will be evaluated. There are also two decision points in the sample acquisition process: which species will be collected for screening, and how samples will be prepared and preserved. Although a number of collecting strategies have been proposed (Perdue 1976; Balick 1990; Cox 1990), no studies have been done that directly compare different methods of sample preparation and preservation and how they affect the discovery rate of new compounds.

Collecting Strategies

Various authors have compared different strategies that could guide collection of samples for screening (Spjut and Perdue 1976; Balick 1990; Cox 1990; Farnsworth 1990). Previously, however, other authors have focused on the potential utility of ethnobotanical information rather than a complete examination of all available strategies. In evaluating various collecting strategies, one must consider how samples will be evaluated, as this is relevant to selecting the appropriate method. Most samples are currently evaluated in high-throughput screens for a single therapeutic area. Hence most labs are set up to efficiently and economically evaluate large numbers of samples for a single therapeutic area rather than a small number of samples for multiple therapeutic areas.

Although Cox (1990) distinguished between random and targeted collecting programs, few, if any, modern programs are truly haphazard efforts to collect plants without knowledge of their relationships. Excluding potential haphazard efforts, a classification of possible collecting strategies is presented in table 5.1. Random collecting efforts can

TABLE 5.1
Classification of Collecting Strategies

Random	Taxonomic	Ecological	Ethnobotanical
Geography Directed	Taxon Directed	Interaction Directed	Random
Classification Directed	Chemical Directed		Healer Directed
			Disease Directed

be either efforts to collect all the available species in a given geographic area or efforts to randomly sample biodiversity across a given classification system. In the first instance, collecting is guided by the selection of geographical location (geography directed); in the second instance, an attempt is made to sample across a given classification system to maximize the biological, hence chemical, diversity. All collecting strategies have certain inherent limitations (Spjut 1985), such as poor representation of small plants or those with scattered populations. The basic division between various directed collecting philosophies is whether they are guided by relationships between plants, the ecological interactions of plants, or use of plants in traditional medicine.

Taxonomic information can be used in several ways to guide collecting programs. Many programs that have been labeled random are, at least theoretically, efforts to obtain a random sample of plant biodiversity, either across a given classification system or in a given geographical region. If a given classification system is robust, it should be predictive (Farris 1979); therefore, maximizing the taxonomic distance between species collected should maximize the chemical diversity evaluated in a particular screen.

Taxonomic information can also be used in other ways to guide collecting. Close relatives of plants that have previously proven to be interesting can be selected based on the assumption that they are most likely to produce similar compounds. This taxon-directed philosophy led to the examination of species related to *Catharanthus roseus* (L.) G. Don (Tin-Wa and Farnsworth 1975) and *Taxus baccata* L. (Elias and Korzenevsky 1992). The distribution of various classes of chemical compounds has frequently been used as a taxonomic marker (Dahlgren 1977), and often it may be possible to predict that a particular class of compounds will prove interesting in a particular screen (Perdue 1976). Thus a chemical-directed philosophy also utilizes and can be based on taxonomic information.

Ecological interactions provide the evolutionary selection pressure that may cause plants to produce the majority of bioactive compounds that they possess. The study of ecological interactions may also be used to direct collecting in a number of ways. Since the diversity of organisms is not evenly distributed, and perhaps more than half the world's species occur in only 7% of the land surface covered by tropical forests (Gentry 1993), ecological interactions may be assumed to be more complex in the tropics, where the number of species with which an individual comes into contact is much greater than in temperate environments. This greater species diversity, and presumed greater ecological interaction, should give rise to greater selection pressure for secondary metabolites (Miller and Brewer 1992). Levin (1976) was able to show that alkaloid diversity was negatively correlated with latitude and that tropical plants were more likely to contain a greater number of alkaloids than those from temperate regions. Thus ecological interactions, hence greater selection pressure for increased diversity of secondary metabolites, should be inversely correlated with latitude; for this reason, tropical plants may be more diverse chemically.

Certain classes of chemical compounds may be present in plants only as a result of specific ecological interactions. Phytoalexins are a class of compounds with interesting therapeutic potential (Barz et al. 1990), but they are produced only in response to physical damage, microbial infection, or insect attack. Collecting programs that ignore or altogether miss plants in this condition will certainly omit a large, interesting group of compounds.

The study of the traditional use of plants as herbal remedies has also been used as a method of selecting plant species for pharmaceutical evaluation (Perdue 1976; Balick 1990; Cox 1990; Farnsworth 1990). Ethnobotanical data can be valuable in several ways. Even in large-scale screening projects, such as the NCI program (Cragg et al. 1994), plants used ethnomedically can be evaluated along with those collected by other methodologies. Because their use in traditional medicine may act as a prescreen, the theoretical rate of discovery of bioactive compounds may be greater than from randomly selected plant samples.

Another way that ethnobotanical data can be used is to evaluate plants against the disease targets for which individual healers use them. Ethnobotanical studies are typically conducted in this manner, and it is an effective method for documenting the ethnobotany of a particular group of people. Unfortunately, the majority of pharmaceutical discovery laboratories are not organized for evaluation of samples against a broad array of targets, but rather focus on a specific disease or therapeutic area. Thus, although this healer-directed approach may be theoretically attractive, it is relatively impractical for final evaluation.

One may also study the use of plants by numerous healers and selectively target collected samples to labs that focus on the therapeutic area for which the plants are used by the healers. Although not a particularly efficient means of preserving ethnobotanical data, it should be an effective means of prescreening plants for pharmaceutical development, and this is the philosophy upon which the company Shaman Pharmaceuticals relies.

Although ethnobotanical data can be very useful, attempts to document that samples collected in this manner result in a higher rate of drug discovery have suffered from several problems. One such study by Balick (1990) examined two sets of plant samples, one collected based on ethnobotanical use and the other randomly selected. The plants were then evaluated in the NCI's HIV screen, and the ethnobotanical samples showed a greater rate of activity than those collected randomly. However, the number of samples was too small to show a statistically significant difference and the results indicated bioactivity only in a primary screen. The percentage of samples that are active in primary screens is always several orders of magnitude greater than the discovery of novel therapeutically useful compounds, primarily because of nonspecific inhibitors and other widespread toxic compounds that lack specific activity and are therefore not useful therapeutically. For this reason, primary screen data is not particularly informative since the majority of active samples are not useful. Cragg et al. (1995) show that less than 1% of the approximately 30,000 extracts tested in the National Cancer Insititute's program have shown selective cytotoxicity, although the rate of positive results in the primary screens is much greater.

One additional observation of ethnobotanical collecting is that it only samples a limited subset of plants. Although it may be a subset with presumed greater bioactivity, there are examples of plants that are not used ethnomedically that have yielded interesting compounds (Thomas and Gereau 1993). In fact, uncommon or rare plants are less likely to be incorporated into an herbal pharmacopoeia, and hence any useful compounds they contain will not be discovered based on ethnobotanically based collection schemes. Uncommon plants may be less likely to be used ethnomedically, because it is less likely that their efficacy will be discovered or because an unreliable or small resource makes their use less attractive to healers.

TABLE 5.2
Distribution of Tanzanian Medicinal Plants

Number of 404 species of Angiosperms with various distributions and their relative percentages				
Tanzania	East Africa	Africa	Africa and Asia	Pantropical
15	67	230	48	44
3.7%	16.6%	56.9%	11.9%	10.9%

Source: Chhabra, Mahunnah, and Mshiu (1987–1993).

Collecting efforts that focus on ethnomedically used plants are likely to consist largely of widespread, common species. As an example, examination of the plants used ethnomedically in eastern Tanzania (Chhabra, Mahunnah, and Mshiu 1987, 1989, 1990a, 1990b, 1991, 1993) shows that only 3.7% are endemic to Tanzania (table 5.2). The vast majority of species in this study were widespread, with 16.6% restricted to East Africa and nearly 23% having ranges extending beyond the continent. Although there are no available estimates for what percentage of the Tanzanian flora is endemic, 25% of the montane flora is endemic (Lovett 1988) and the 3.7% of ethnomedically used plants is clearly below the national average.

Although each of the philosophies discussed above has inherent strengths and weaknesses, other, more practical considerations need to be weighed. Spjut (1985) has pointed out that regardless of the collecting philosophy, certain plant types are likely to be poorly represented in any large screen. These include small plants or sparsely distributed plants, which are difficult to gather in sufficient quantity for evaluation. Spjut listed six large families (Aizoaceae, Araceae, Bromeliaceae, Cyperaceae, Orchidaceae, and Poaceae), which he estimated to contain more than 35,000 species, that were very poorly represented in the NCI program.

An additional consideration is that all the collecting philosophies in table 5.1 are based on deciding which sample should be included. As a program proceeds a second level of decision making is possible: a group may be excluded from further evaluation based on poor initial results or a promising group can become the focus of increased attention.

It should be clear from this discussion that there is no single optimal approach to sample gathering but that success may depend on utilizing aspects of several of these ideologies. The actual choice of which guidelines to use is also dependent on how the samples will be evaluated. Thus evaluation of a species used by many groups of healers for the same therapeutic problem is an efficient choice for a lab that focuses on a single disease, whereas the random evaluation in a single screen of ethnobotanical samples used for a variety of ailments is not as likely to be efficient.

Sample Preservation Methodology

Almost always a period of time elapses between the collection of a sample and its evaluation for pharmaceutical potential in the laboratory. Little attention, however, has been given to selection of the optimal method of handling and preserving the samples. The current literature contains no direct comparison of rates of discovery of novel, therapeutically useful compounds from samples prepared by different methods, and this is a very difficult question to test directly.

The ideal experiment to answer this question would be to collect plant material for screening and prepare and preserve replicate samples by a number of different methods. Each of the samples would then be screened for novel, therapeutically useful compounds, and the rates of discovery could be compared to determine which sample preservation technique was most effective. However, given that expected rates of discovery are predicted to be low, on the order of 1 per 50,000 samples in the National Cancer Institute's cancer screens (Cragg et al. 1995), it would be necessary to screen an almost infinite number of samples in order to obtain a statistically significant difference in the discovery rates of the various preservation methods.

Although direct comparison of sample preservation techniques has not been done and would be difficult to do, there are data that indirectly address this question. Both qualitative and quantitative differences in how certain classes of compounds persist in samples prepared in different ways has been studied, and examination of many of these studies shows that a number of trends with predictive value exist.

Several studies have shown that the quality of extractable DNA from dried leaves varies according to the preservation method. Pyle and Adams (1989) were able to demonstrate that good yields of genomic DNA were possible from fresh, frozen, and dried leaves but that high molecular weight DNA was rapidly denatured by all the chemical preservation methods that they evaluated. The quality of extractable spinach DNA also varied according to the temperature at which the samples were stored (Adams, Do, and Ge-lin 1991). Several other studies, however, have shown that the quality of extractable DNA does not vary between samples dried with contact paper, corrugates, or silica gel (Harris 1993; Harris and Robinson 1994).

Perhaps more directly relevant to a discussion of how to best preserve plant samples for later pharmaceutical evaluation is the research of Jacquin-Dubreuil et al. (1989), who studied the effects of various drying methods on the qualitative and quantitative preservation of the alkaloids nicotine and nor-nicotine. If these two may be assumed to be representative of the class of compounds in general, this can be particularly informative as alkaloids are often therapeutically useful. Comparison of the drying methods used in all these studies indicated that the more rapidly samples were dried, without exposing them to extreme conditions, the better the preservation of the samples' chemical integrity. This probably indicates that the best preservation occurs by stopping cell activity, hence enzymatic degradation, yet at the same time not exposing the sample to extreme heat or other factors that would cause degradation of chemicals.

Although little data are available to compare many of the possible methods of sample preservation, it is clear that variation occurs. Furthermore, data that are available allow speculation about various methods of drying. In many regions of the world where collecting is likely to take place, drying may be the only preservation method that is logistically possible, and thus the most important to study carefully. Current studies indicate that material is best preserved by drying if the material is left in relatively large pieces, rather than ground or chopped into small pieces (D. Seigler, pers. comm.), if it is dried quickly but without exposure to extreme temperatures, and if it is dried without direct exposure to sunlight, which may cause photo-oxidation of active components. Further data from Adams, Do, and Ge-lin (1991) indicate that, once samples are dry, storage at cold temperatures will aid in the preservation of high molecular weight DNA and probably other chemical constituents as well.

Acknowledgments

I thank C. Dietrich and A. Randrianasolo for bibliographic assistance and D. K. Harder and T. F. Stuessy for comments on the manuscript. National Cancer Institute contract NCI-CM-17515–30 provided financial support and experience with much of the subject matter, and their support is gratefully acknowledged.

Literature Cited

Adams, R. P., N. Do, and C. Ge-lin. 1991. Preservation of DNA in plant specimens from tropical species by desiccation. In R. P. Adams and J. E. Adams, eds., *Conservation of Plant Genes: DNA Banking and in Vitro Biotechnology*, pp. 135–152. San Diego: Academic Press.

Balandrin, M. F., A. D. Kinghorn, and N. R. Farnsworth. 1993. Plant-derived natural products in drug discovery and development. In A. D. Kinghorn and M. F. Balandrin, eds., *Human Medicinal Agents from Plants*, pp. 2–12. Washington, D.C.: American Chemical Society.

Balick, M. J. 1990. Ethnobotany and the identification of therapeutic agents from the rainforest. In D. J. Chadwick and J. Marsh, eds., *Bioactive Compounds from Plants*, pp. 22–39. Ciba Foundation Symposium 154. Chichester: Wiley.

Barz, W., W. Bless, G. Borger-Papendorf, W. Gunia, U. Mackenbrock, D. Meier, Ch. Otto, and E. Super. 1990. Phytoalexins as part of induced defence reactions in plants: Their elicitation, function, and metabolism. In D. J. Chadwick and J. Marsh, eds., *Bioactive Compounds from Plants*, pp. 140–156. Ciba Foundation Symposium 154. Chichester: Wiley.

Bugg, C. E., W. M. Carson, and J. A. Montgomery. 1993. Drugs by design. *Sci. Amer.* 269:92–98.

Chhabra, S. C., R.L.A. Mahunah, and E. N. Mshiu. 1987. Plants used in traditional medicine in Eastern Tanzania. Part 1. Pteridophytes and Angiosperms (Acanthaceae to Canellaceae). *J. Ethnopharmacology* 21:253–277.

——. 1989. Plants used in traditional medicine in Eastern Tanzania. Part 2. Angiosperms (Capparidaceae to Ebenaceae). *J. Ethnopharmacology* 25:339–359.

——. 1990a. Plants used in traditional medicine in Eastern Tanzania. Part 3. Angiosperms (Euphorbiaceae to Menispermaceae). *J. Ethnopharmacology* 28:255–283.

——. 1990b. Plants used in traditional medicine in Eastern Tanzania. Part 4. Angiosperms (Mimosaceae to Papilionaceae). *J. Ethnopharmacology* 29:295–323.

——. 1991. Plants used in traditional medicine in Eastern Tanzania. Part 5. Angiosperms (Passifloraceae to Sapindaceae). *J. Ethnopharmacology* 33:143–157.

——. 1993. Plants used in traditional medicine in Eastern Tanzania. Part 6. Angiosperms (Sapotaceae to Zingiberaceae). *J. Ethnopharmacology* 39:83–103.

Cox, P. A. 1990. Ethnopharmacology and the search for new drugs. In D. J. Chadwick and J. Marsh, eds., *Bioactive Compounds from Plants*, pp. 40–55. Ciba Foundation Symposium 154. Chichester: Wiley.

Cragg, G. M., M. R. Boyd, M. R. Grever, T. D. Mays, D. J. Newman, and S. A. Schepartz. 1994. Natural product drug discovery and development at the National Cancer Institute: Policies for international collaboration and compensation. In R. P. Adams, J. S. Miller, E. M. Golenberg, and J. E. Adams, eds., *Conservation of Plant Genes II: Utilization of Ancient and Modern DNA*, pp.

221–232. Monogr. Syst. Bot. Missouri Bot. Gard. No. 48. St. Louis: Missouri Botanical Garden.

Cragg, G. M., M. R. Boyd, M. R. Grever, and S. A. Schepartz. 1995. Pharmaceutical prospecting and the potential for pharmaceutical crops. Natural product drug discovery and development at the United States National Cancer Institute. *Ann. Missouri Bot. Gard.* 82:47–53.

Dahlgren, R. 1977. A commentary of the diagrammatic presentation of the angiosperms in relation to the distribution of character states. *Plant Syst. Evol. Suppl.* 1:253–283.

Decosterd, L. A., I. C. Parsons, K. R. Gustafson, J. H. Cardellina II, J. B. McMahon, G. M. Cragg, Y. Murata, L. K. Pannel, J. R. Steiner, J. Clardy, and M. R. Boyd. 1993. Structure, absolute stereochemistry, and synthesis of conocurvone, a potent, novel HIV-inhibitory naphthoquinone trimer from a *Conospermum* sp. *J. Amer. Chem. Soc.* 115:6673–6679.

Elias, T. S. and V. V. Korzenevsky. 1992. The presence of taxol and related compounds in *Taxus baccata* native to the Ukraine (Crimea), Georgia, and southern Russia. *Aliso* 13:463–470.

Farnsworth, N. R. 1990. The role of ethnopharmacology in drug development. In D. J. Chadwick and J. Marsh, eds., *Bioactive Compounds from Plants*, pp. 2–21. Ciba Foundation Symposium 154. Chichester: Wiley.

Farnsworth, N. R. and D. D. Soejarto. 1985. Potential consequence of plant extinction in the United States on the current and future availability of prescription drugs. *Econ. Bot.* 39:231–240.

Farris, J. S. 1979. The information content of the phylogenetic system. *Syst. Zool.* 28:483–519.

Gentry, A. H. 1993. Tropical forest biodiversity and the potential for new medicinal plants. In A. D. Kinghorn and M. F. Balandrin, eds., *Human Medicinal Agents*, pp. 13–24. Washington, D.C.: American Chemical Society.

Gustafson, K. R., J. H. Cardellina II, J. B. McMahon, R. J. Gulakowski, J. Ishitoya, Z. Szallasi, N. E. Lewin, P. M. Blumberg, O. S. Weislow, J. A. Beutler, R. W. Buckheit, Jr., G. M. Cragg, P. A. Cox, J. P. Bader, and M. R. Boyd. 1992. A nonpromoting phorbol from the Samoan medicinal plant *Homalanthus nutans* inhibits cell killing by HIV-1. *J. Med. Chem.* 35:1978–1986.

Harris, S. A. 1993. DNA analysis of tropical plant species: An assessment of different drying methods. *Pl. Syst. Evol.* 188:57–64.

Harris, S. A. and J. Robinson. 1994. Preservation of tropical plant material for molecular analyses. In Adams et al., eds., *Conservation of Plant Genes II*, pp. 83–92. Monogr. Syst. Bot. Missouri Bot. Gard. No. 48. St. Louis: Missouri Botanical Garden.

Jacquin-Dubreuil, A., C. Breda, M. Lescot-Layer, and L. Allorge-Boiteau. 1989. Comparison of the effects of a microwave drying method to currently used methods on the retention of morphological and chemical leaf characters in *Nicotiana tabacum* L. cv. *Samsun*. *Taxon* 38:591–596.

Kashman, Y., K. R. Gustafson, R. W. Fuller, J. H. Cardellina II, J. B. McMahon, M.

J. Currens, R. W. Buckheit, S. H. Hughs, G. M. Cragg, and M. R. Boyd. 1992. The Calanolides, a novel HIV-inhibitory class of coumarin derivatives from the rainforest tree, *Calophyllum lanigerum*. *J. Med. Chem.* 35:2735–2743.

Kingston, D. G .I. 1993. Taxol, an exciting anticancer drug from *Taxus brevifolia*. In A. D. Kinghorn and M. F. Balandrin, eds., *Human Medicinal Agents*, pp. 138–148. Washington, D.C.: American Chemical Society.

Klayman, D. L. 1993. *Artemisia annua* from weed to respectable antimalarial plant. In A. D. Kinghorn and M. F. Balandrin, eds., *Human Medicinal Agents*, pp. 242–255. Washington, D.C.: American Chemical Society.

Levin, D. A. 1976. Alkaloid-bearing plants in ecogeographic perspective. *Amer. Nat.* 110:261–284.

Lovett, J. C. 1988. Endemism and affinities of the Tanzanian montane flora. In P. Goldblatt and P. P. Lowry, eds., *Modern Systematic Studies in African Botany*, pp. 591–598. Monogr. Syst. Bot. Missouri Bot. Gard. No. 25. St. Louis: Missouri Botanical Garden.

Manfredi, K. P., J. W. Blunt, J. H. Cardellina II, J. B. McMahon, L. K. Pannel, G. M. Cragg, and M. R. Boyd. 1991. Novel alkaloids from the tropical plant *Ancistrocladus abbreviatus* inhibit cell-killing by HIV-1 and HIV-2. *J. Med. Chem.* 34:3402–3405.

Meyer, S. L. 1946. The war against malaria. *J. Tenn. Acad. Sci.* 21:346–350.

Miller, J. S. and S. J. Brewer. 1992. The discovery of medicines and forest conservation. In R. P. Adams and J. E. Adams, eds., *Conservation of Plant Genes: DNA Banking and in Vitro Biotechnology*, pp. 119–134. San Diego: Academic Press.

Miller, J. S. and D. K. Harder. 1994. Models for ethical collaboration in biodiversity prospecting. In Adams et al., *Conservation of Plant Genes II*, pp. 239–243.

Noble, R. L., C. T. Beer, and J. H. Cutts. 1958. Role of chance observations in chemotherapy: *Vinca rosea*. *Ann. N.Y. Acad. Sci.* 76:882–894.

Perdue, R. E. 1976. Procurement of plant materials for antitumor screening. *Cancer Treat. Rep.* 60:987–998.

Population Reference Bureau. 1993. World Population Data Sheet. Washington, D.C.: Population Reference Bureau.

Pyle, M. M. and R. P. Adams. 1989. In situ preservation of DNA in plant specimens. *Taxon* 38:576–581.

Spjut, R. W. 1985. Limitations of a random screen: Search for new anticancer drugs in higher plants. *Econ. Bot.* 39:266–288.

Spjut, R. W. and R. E. Perdue. 1976. Plant folklore: A tool for predicting sources of antitumor activity. *Cancer Treatment Reports* 60:979–985.

Svoboda, G. H. 1961. Alkaloids of *Vinca rosea* (*Catharanthus roseus*). Part 9. Extraction and characterization of leurosidine and leurocristine. *Lloydia* 24:173–178.

Thomas, D. W. and R. E. Gereau. 1993. *Ancistrocladus korupensis* (Ancistrocladaceae): A new species of liana from Cameroon. *Novon* 3:494–498.

Tin-Wa, M. and N. R. Farnsworth. 1975. The phytochemistry of the minor *Catharanthus* species. In W. I. Taylor and N. R. Farnsworth, eds., *The Catharanthus Alkaloids*, pp. 85–124. New York: Marcel Dekker.

Vagelos, R. 1991. Are prescription drug prices high? *Science* 252:1080–1084.

Wani, M. C., H. L. Taylor, M. E. Wall, P. Coggon, and A. T. McPhail. 1971. Plant antitumor agents. Part 6. The isolation and structure of Taxol, a novel antileukemic and antitumor agent from *Taxus brevifolia. J. Amer. Chem. Soc.* 93:2325–2327.

Webster, T. D., R. Ramirez-Solis, R. D. Bies, D. L. Nelson, L. Corbo, and C. T. Caskey. 1989. The human genome project and its impact on medicine. In J. G. Fortner and J. E. Rhoads, eds., *Accomplishments in Cancer Research 1989*, pp. 276–295. General Motors Cancer Research Foundation. Philadelphia: J. B. Lippincott.

PART 3

Collection of Images of Plant Diversity

Sometimes it is simply not possible to obtain actual plant specimens. Plant diversity is destroyed in some areas faster than it can be collected. In view of the insufficient statistical profile that the final sample will represent (Baum, chapter 4), we might ask if images of plant diversity might somehow fill in the gaps, and if so, how should this be done? At one level, we have all used photographs and illustrations in our own research and teaching. James White in chapter 6 touches on issues of maintaining and cataloging botanical art works and their history from the Age of the Herbalists (1460–1660) to the present. He shows that illustrations provide indispensable additions to photographs and, in some cases, more easily reveal diagnostic features. One challenge in this area facing our community is for better indexes so that we can more effectively use the many illustrations of plants that already exist. White makes the important point that these images must also be tied to voucher specimens. Without this connection, it is most difficult to evaluate illustrations fully, and they lose proportionate research value. They may still retain high aesthetic value, however.

Part of the problem in dealing with illustrations has been the lack of means to publish them in more convenient and inexpensive formats. All of us might wish to have a collection of 20,000 botanical prints and paintings, but even in a series of colored reproductions published as books, this would be large and expensive. Newer technologies of CD-ROM and videodisc now make these projects feasible, as pointed out by David Kramer in chapter 7. A further step, which is not yet economically feasible, is for making all images available on-line for computer access worldwide. It requires no clairvoyance to believe that this will happen in the near future, and the project presented by Kramer on establishing a Visual Archive of the Flora of Ohio is just one interesting example of what might be accomplished.

An even more mind-stretching contribution is by Don Stredney (chapter 8), who tells us about the potential of virtual reality for documenting phytodiversity. In his symposium presentation, he showed several videos that revealed, in a most spectacular fashion, the potential of these techniques for depicting plant form and structure. We are quite certain that it will be possible in the future to input available data about plants and reconstruct them in a virtual environment for education and research. The potentials are intriguing, astounding, and somewhat frightening. Although Stredney might disagree, to us there appear some real dangers for human society when our virtual world approximates the real one. Like it or not, that time has arrived and will be an even greater force in the future. The teaching potentials are enormous; prototypes already exist for instructing surgeons in surgical techniques with virtual situations that allow them to make incisions and guide the course of maneuvers as practice for real operations with real patients. Pushed to the extreme, virtual reality will require us to decide in what type of world we wish to live in the future. There is much to ponder here.

Michael DeMers in chapter 9 tells us what most of us already know, that the age of GIS (Geographic Information Systems) is upon us. We simply have to learn more about it, get linked into it, and begin to use it routinely. Need for more precise locality and environmental data will drive acquisition of satellite and remote sensing information plus "ground truth" from field work, so essential for data verification. DeMers presents an excellent overview of the potentials, and some of the difficulties, with establishing and using a GIS system in the context of vegetation analysis.

6

Images of Plant Diversity: Virtual Specimens and Graphic Glosses

JAMES J. WHITE

Reproductions of botanical art and illustrations are an integral part of the literature of botany and horticulture. Images are used in botany not just as glosses that enhance understanding, but often as virtual specimens. Botany records its findings not only in words and numbers but in pictures, and these images are not just an important aesthetic heritage; they remain a valuable scientific resource.

Among the first herbaria I ever visited were those at the U.S. National Arboretum and the Smithsonian Institution. At the time I thought it curious and wonderful to see in the former some of the more than 7,000 watercolors of fruits—many of them the nomenclatural types of the cultivars depicted—and in the latter the brush and ink drawings of northwestern conifers by Frederick A. Walpole, turn-of-the-century United States Department of Agriculture (USDA) and Smithsonian Institution artist. These artists had left botanical records that also were works of art.

Fosberg and Sachet (1965:18) have called the modern herbarium "a great filing system for information about plants, both primary in the form of actual specimens of the plants and secondary in the form of published information, pictures, and recorded notes." They added that "this system is ideal also for the inclusion of notes, photographs, drawings, plates, and clippings from published literature, which are mounted on similar sheets of paper and filed under their proper family, genus, or species."

"Systematics collections of plants and animals are the only permanent record of the earth's biota"—so begins *America's Systematics Collections: A National Plan* (Irwin et al. 1973:1)—"and the specialized libraries attached to these collections are the written record of the earth's natural history." I propose making one addition to this statement so that it reads, "Systematics collections of plants and animals, along with the images and literature collectively, are the only permanent record of the earth's biota, and the specialized libraries attached to these collections are the written record of the earth's natural history."

"Library holdings provide an essential resource for research in systematic biology," the *Plan* continues (Irwin et al. 1973:15). "The specialized systematics libraries that have grown with collections and research are not limited to bound books and periodicals; they may also include a variety of related aids: various card indices, catalogues, manuscripts, illustrations, microfiche records of historically important herbaria, cartographic information, and bibliographical and biographical files." Images are a crucial element in recording our plant biodiversity.

History of Botanical Images

Some of the earliest illustrations of plants were made for the searchers of herbs. Then the useful yielded to the beautiful as flowers in seventeenth-century gardens of wealthy patrons were drawn. Later, as scientific botany flourished, all plants began to serve as models for the artist.

According to W. T. Stearn (1993:283), "The progress of biology from the 16th century onwards has been closely linked to the publication of illustrations enabling individual species to be distinguished and thereafter designated by acceptable names." Or in the words of my predecessor, John Brindle (1964:148–149), the Hunt Institute's first curator of art:

> Botanical science provides a good example of the interaction between scientific illustration and progress. The wide distribution of books containing printed representations of plates, based on direct observation,

conveyed exact knowledge to an extent never before possible, and gave strong impetus to the study and development of the many classification systems that arose during the eighteenth and nineteenth centuries.

The primitive state of botany in the Middle Ages is clearly reflected in the innumerable plant illustrations to be seen in early herbals, illustrations which had their remote origins in the naturalistic paintings of classical antiquity, as transmitted through the monumental *De materia medica* of Dioscorides, a work of the first century A.D. which survives in the famous sixth-century manuscript known as the *Codex Aniciae Julianae* in Vienna.

The process of copying and recopying, without resource to direct observation of the plants themselves, continued throughout the centuries of the Dark Ages and of the Middle Ages with deadening effect, until images of the plants, as seen, for instance, in some of the early printed herbals (e.g., *Hortus sanitatis*), were mere decorative embellishments of the text with little usefulness for study or identification.

With the sixteenth-century development of book printing, and of wood-cut illustrations, a fresh start in plant observation and representation was made. The pivotal position is held by Otto Brunfels's *Herbarum vivae eicones* (1530) illustrated with vigorous woodcuts by Hans Weiditz, an associate of Dürer. Weiditz took his inspiration from Nature, used his own eyes and drew with great precision, setting a new high standard for plant illustration.

Subsequent progress through the following centuries was accompanied and spurred forward by the proliferation of print-making processes. It is improbable that this progress could have been so steady or so rapid without the means of disseminating knowledge by graphic representation.

Those paintings and drawings that are recognized as the originals on which particular printed plates are based are especially valuable for purposes of comparison. Redouté's watercolors afford several examples, and it can be seen to what degree his intent and style are realized in the color-printed and hand-retouched stipple engravings of *Les roses*, for instance. But the comparison is instructive, whether we consider work from the early nineteenth century or paintings and drawings reproduced by twentieth-century photomechanical processes.

Since they have an intrinsic value, the original artwork that has been reproduced is valuable but the notations by artists, botanists, and printers are also valuable. One of my favorite scribblings in our collection is on an artist's proof engraving of *Diclidocarpus* for the "U.S. Exploring Expedition": "This is an old plate which was rejected, & the principal part used on another plate. Destroy it! A. Gray."

Images at the Hunt Institute

Artwork (and a small collection of photographs) has been an important part of the Hunt Institute since its founding in the early 1960s. Botanical imagery from the Renaissance on is well represented in the institute's collection of more than 30,000 watercolors, drawings, and original prints—a major resource for the history of botanical art and illustration. In addition, our library contains 20,000 books, many of them illustrated ones from the 1730–1840 period.

In both acquisitions and program, the department has pursued a specialty in twentieth-century works and artists, and our coverage in this field is now unmatched anywhere. We solicit and maintain contacts worldwide with artists and illustrators who have treated plant subjects at professional levels of achievement, and our series of International Exhibitions of Botanical Art and Illustration is considered the premier showcase for contemporary artists working in the genre. The cumulative index of the latest International catalogue now lists more than 650 such artists from around the world. Selections from the most recent International are included in a ready-to-hang travel show that is circulated at modest rental fees to museums, schools, botanical gardens, and other institutions throughout the United States.

Special parts of the art collection include the Torner collection of the almost 2,000 botanical and zoological illustrations made during the Spanish royal expedition of 1787–1803 to New Spain; the Hitchcock-Chase collection of more than 4,700 ink drawings of grasses, assembled by the Smithsonian agrostologists Albert S. Hitchcock (1865–1935) and Agnes Chase (1869–1963); and the USDA Forest Service collection of more than 2,800 ink drawings of woody species. The institute also has an important collection of drawings by Frederick A. Walpole, late nineteenth-century American artist, associated with the USDA and the Smithsonian Institution.

Copies of the institute's artwork are available for reproduction, depending on copyright restrictions and intended uses.

Art Catalogue

The first part of the catalogue of the botanical art collection at the Hunt Institute (White and Smith 1985) contains a description of the catalogue entries. Already complete in machine-readable form, the database is being published by parts, and we hope eventually to record the entire collection in full color on CD-ROM. Besides archiving the collection photographically, the CD-ROM will facilitate such departmental services as identifying artists and sources of botanical illustrations, furnishing images to authors, publishers, and the general public, and lending works for exhibition. Together, the catalogue and CD-ROM will be a resource of significant value in the plant sciences and related fields such as conservation, ecology, agronomy, horticulture, and biological illustration.

Organization of the Hunt Art Collection

The paintings, drawings, and prints of the Hunt collection are organized to facilitate storage, display, and study under sound conservation conditions. Since the material varies greatly in size, it is divided for storage purposes into six size categories. Items in smaller categories are stored in envelopes consisting of a clear mylar facing and a backing of all-cotton rag mounting board, sealed on three sides with double-sided 3-M Scotch Brand tape. Materials in other categories are matted according to standard museum practice, that is, they are hinged with Japanese paper tabs and wheat starch paste and are faced by a mat cut from cotton rag mat board with apertures to accommodate the subject. All material is stored at a temperature of 68–70°F and a relative humidity of 50–55 percent. Paintings, drawings, and prints are exhibited as needed in their storage mats in metal frames, matched in size to accommodate the standard mats. Display material is thus readily interchanged, and one exhibition may be quickly replaced by another.

Original Botanical Art Project

The institute is developing a research database to be available on the Internet and published eventually in hardcopy on original botanical paintings and drawings (not prints or sculpture) from any time period done in traditional media such as watercolor, pastel, ink, or pencil (but not oil) in public and private collections. We are interested in scientific illustrations of plants or any artwork that accurately depicts botanical subjects as their main subjects and may be of taxonomic importance.

Information about still-life and impressionistic and traditional oriental paintings is of secondary importance for this survey.

Artwork and Photography

C. P. Smyth (1860), in the *Transactions of the Edinburgh Botanical Society*, insisted that artists had grossly misrepresented the growth habit of Dragon trees of Teneriffe ("Dracæna Draco") and published a photograph to support his contention that only photography could preserve reliable records of such natural phenomena. In his view, artists tended to translate nature into the conventionalized images that European artistic training demanded, rather than to satisfy the objective needs of science. Smyth probably overstated his general argument, even though this particular case was proven. He also demonstrated to the Edinburgh Botanical Society that a significant new medium for botanical illustration had arrived. However, that society and most others continued to employ artists to prepare the great bulk of scientific botanical illustrations (Schaaf 1981; Bridson and Wendel 1986:143).

Successful methods of producing photographic illustrations in printer's ink came into widespread use only in the early 1880s, forty years after photography first came to public notice. Photography remains valuable for recording growth phases, seasonal changes, environmental associations, and so on.

Wilfrid Blunt, in his classic *The Art of Botanical Illustration* (1950:1–2), wrote: "In the present century, the photographer has largely usurped the place of the botanical artist, though happily there is still a certain amount of work in which the camera is no adequate substitute for the pencil." Forty-three years after Blunt's comment, the images connected with botany still are mostly artwork. Though photography has entered the scene in recent times, it probably won't replace art (Fig. 6.1). That natural history art is thriving is attested to by the growing membership and activities of the Guild of Natural Science Illustrators, based at the Smithsonian in Washington but international in scope.

Voucher Specimens

For full scientific legitimacy all images should be vouchered if they are to serve as virtual specimens. In artwork one usually can see what one

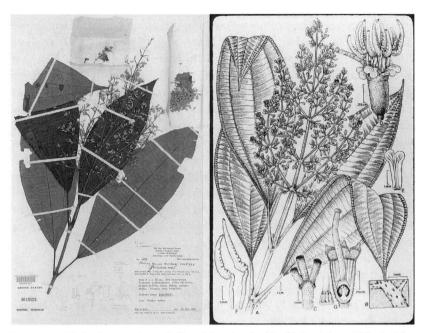

FIGURE 6.1 Herbarium specimen (*left*) and drawing (*right*) of *Miconia boomii* Wurdack by Alice Tangerini, Department of Botany, Smithsonian Institution, Washington, D.C. (from Wurdack 1988:11).

needs for identification, but the majority of photos require voucher specimens to confirm what they represent. This is a problem for anyone organizing archives of photos of plants. It is a shame that so many botanists have good pictures of plants but ones that are not organized or readily retrievable and that leave us uncertain as to what they actually represent. Some botanists (like Tod Stuessy) have made a habit of vouchering photos. He includes a collection number with each photo. This is what must be done. Even if such images are not used as virtual specimens, they certainly are graphic glosses that help with identifications. In either case, one does not want to be in a position of providing photos of subjects the identities of which cannot be confirmed.

Go to a library to identify a plant. How disappointing it is not to find an illustration—and not because we are lazy and like to look at pictures. A friend of mine who acquired a field guide on identifying birds (with photos) has replaced it with one containing artwork depicting birds. Floras and field guides are less useful when identifying characteristics of the plants and critters cannot be simply and clearly outlined. Photographs focus on part of the organism and contain a lot of visual noise. *The Guild Handbook of Scientific Illustration* (Hodges 1989) reports, "Although photography plays a very important role in the collection of data in the field, scientific illustrations are often superior to photographs for recording visual information. Good illustrations will eliminate confusing shadows, highlight diagnostic features, separate and show clearly objects that are hidden or inaccessible in nature." Few scientists are good photographers; however, botanist Sherwin Carlquist seems to be an exception. His careful composition and use of a Hasselblad camera have produced extraordinary results that eventually may be reproduced by the Hunt Institute on CD-ROM.

One picture is worth ten thousand words, say the Chinese. If systematics is a field concerned with diversity, then clearly pictures are an important resource for the botanist.

Literature Cited

Blunt, W. 1950. *The Art of Botanical Illustration*. London: Collins.

Bridson, G.D.R. and D. E. Wendel. 1986. *Printmaking in the Service of Botany*. Pittsburgh: Hunt Institute.

Brindle, J. V. 1964. Program and organization of the collection. *Huntia* 1: 148–155.

Fosberg, F. R. and M. H. Sachet. 1965. Manual for tropical herbaria. *Reg. veg.* 39: 1–132.

Hodges, E.R.S., ed. 1989. *The Guild Handbook of Scientific Illustration*. New York: Van Nostrand Reinhold.

Irwin, H. S., W. W. Payne, D. M. Bates, and P. S. Humphrey, comps. and eds. 1973. *America's Systematics Collections: A National Plan*. Lawrence, Kansas: Association of Systematics Collections.

Schaaf, L. 1981. Piazzi Smyth at Teneriffe. Part 2: Photography and the disciples of Constable and Harding. *History of Photography* 5: 27–50.

Smyth, C. P. 1860. On the manner of growth of Dracæna Draco in its natural habitat, as illustrating some disputed points in vegetable physiology. *Transactions of the Edinburgh Botanical Society* 11: 250–261.

Stearn, W. T. 1993. Reviews of publications. *Botanical Journal of the Linnean Society* 112: 283.

White, J. J., with the assistance of Elizabeth R. Smith. 1985. *Catalogue of the Botanical Art Collection at the Hunt Institute*. Part 1: *Plant Portraits, Artists A-D*. Pittsburgh: Hunt Institute.

Wurdack, J. 1988. New Melastomataceae from Peru and Bolivia. *Brittonia* 40: 11.

7

The Use of New Technologies to Create a Visual Archive of Plant Diversity

DAVID W. KRAMER

Plant biologists recognize the urgency to record, study, and preserve species that are threatened by the alarming loss of habitat around the world. Herbarium specimens and liquid-preserved specimens are essential for scientific research and will continue to be important in the foreseeable future. But these traditional forms of preservation must be supplemented with a visual archive of still photographs, drawings, and movies, just as they have been in the past. Visual images are essential for publication of scientific research and just as important for education. Scientists must recognize that research will provide us with information about biodiversity, but threatened species will be conserved only when an informed electorate, their governments, and their private conservation organizations recognize the problem and are willing to take action to slow or halt the loss of habitat with its concomitant loss of biodiversity.

New technologies such as videodisc and CD-ROM must be considered for both research and pedagogical purposes. In spite of serious shortcomings of current digital technology, digitization appears to be the most promising technology for the future. Digitized images can be stored and manipulated in computerized databases and can be transmitted through computer networks locally and globally via Internet. Pilot projects to test the efficacy of digitized visual archives should be launched now. There are many decisions to be made in the early planning stages of these projects so that the image archive will meet the needs of research scientists, teachers, and learners.

Biologists (Kim and Knutson 1986; Wilson and Peter 1988; and Wilson 1992) and legislators (U.S. Congress, Office of Technology Assessment 1992) have been sounding the alarm for decades about the tragic loss of biodiversity through habitat destruction. In spite of these warnings some have admitted that the full extent of the loss of biodiversity is impossible to predict because "tens of millions of species remain unknown to us" (Anonymous 1994:1). Recent arguments for a national biological survey (Kim and Knutson 1986) and even a global biological survey (Anonymous 1994) make this an ideal time to examine the techniques that should be employed in recording biodiversity. Here we are focusing on documenting the hundreds of thousands of plant species that inhabit our biosphere—techniques for sampling the green world.

The "tried and true" traditional methods of collecting and preserving plants with herbarium specimens and liquid-preserved specimens will have their place in this frantic effort to record diversity, but we must also consider how newer technologies can support not only the gathering and storing of knowledge but its subsequent dissemination as well.

Our first consideration must be to identify those techniques that facilitate the work of the systematists, ecologists, and other biologists who have the skills to identify and name new species and explain the interactions of organisms in an ecosystem. However, if we are concerned about conserving species, not merely naming and listing them, we must admit that scientists alone cannot reverse the trends. Conservation of habitats and species ultimately is a matter of public policy which can be adopted only with widespread public support. Biologists, therefore, have an obligation not only to document biodiversity for themselves but to educate the public so that enlightened conservation policies are adopted.

Only a few of the formal scientific discussions of the conservation of biodiversity have recognized the need to educate the public. Perhaps

Soulé (1986:11) stated it most succinctly: "It won't be long before many conservation biologists are spending more time at community meetings than in the field or laboratory." The Paris Declaration of the Tenth World Forestry Congress (Unasylva 1992) called on decision makers to (among other things) "raise the awareness of the public, and more particularly of young generations, and disseminate information on forest issues so they will be better appreciated by all people." Lucas (1993) also pointed to the need for educating the general public at a time when knowledge about plants is declining.

Traditional Methods of Documenting Plant Diversity

Few plant biologists would dispute the value of herbarium specimens (and liquid-preserved specimens) as a means of preserving a record of the flora of the world. Herbarium specimens are not only to examine, catalog, and study, but they also are sources of research material for plant anatomists, cytologists, and, more recently, molecular geneticists. It is easy to understand that the removal of parts from herbarium specimens has long been a source of irritation with herbarium curators who are only partially appeased by the requisite annotation label (Stern and Chambers 1960). Yet it can be successfully argued that the detailed study of these detached parts, often the only readily available material of some species, has added significantly to our understanding of a large number of taxa.

There is a long history, predating the "Information Highway" by hundreds of years, of herbarium specimens being shipped to very remote herbaria and research laboratories around the globe for study by professional biologists. Clearly herbarium specimens are very useful to plant biologists, and it is unlikely we will replace them entirely with any of the new technologies, now or in the future.

Nevertheless it is useful to ask ourselves if herbarium specimens adequately serve all purposes. When systematists correspond with one another through informal communication or through formal publication, they rely on line drawings or photographs to document the differences between species. Most systematists are also teachers and, although we use herbarium specimens in our courses on taxonomy and systematics, those specimens most often are drawn from a special "teaching collection," not from the "real" herbarium. This is because herbarium specimens are fragile and can withstand only limited exposure to students who rarely handle the sheets the way they have been instructed.

As teachers and researchers we rely on more durable visual evidence: drawings, 35-mm slides, and photographs. Visual archives of plants in any of the usual forms are difficult to maintain. First there is the trouble and expense of making the drawing or taking the photograph and having it developed. Then there is the question of cataloging and storing in some way that allows easy retrieval. When we are ready to use the images, we must select the ones we want, remove them from storage, place the drawings on display and the slides in a slide tray, and then, when we are finished, reverse the process. Acquiring a set of images in support of even the most elementary course can be a life's work.

I have been presenting workshops for high school biology teachers during the last three summers. This experience has convinced me that there is a pressing need for a visual archive of our flora that is complete, of high quality, inexpensive, and easier to manage than print or slide collections. If those of us who teach in the colleges and universities can barely find the time or resources for slides and slide management, think of our impoverished colleagues in the elementary and secondary schools.

While speaking of teachers we must remember those who are engaged in "informal education" in our libraries, museums, and state agencies like the Department of Natural Resources. They, too, rely heavily on photos of plants to carry out their service, education, and research missions and could benefit from an inexpensive, durable, and reasonably complete visual archive.

Using Newer Technologies

Now and certainly for the future we must consider the possibility of storing photos, drawings, and text information in computers or in forms that can be manipulated by computers. By so doing the images can be transported, almost instantaneously, around the world via Internet. Finally we have the tools to store a collection of outstanding slides, drawings, text, even movies that could enliven the dullest of botany lessons. But we would have even more—systematists and students could have direct access to a nearly indestructible archive of images that they could use for research projects and then could "cut and paste" into their class assignments and publications. Teachers at all levels could use the images to support lectures or embed them in interactive computer software for their students. Let's look at the storage possibilities.

Videodiscs

Videodiscs, also called "laser discs," are capable of storing 54,000 images (e.g., 54,000 35-mm slide images) on *each side* of a standard videodisc. To put this in perspective, just *one side* is the equivalent of 675 carousel trays (80 slides per tray)! When *both* sides are recorded, each disc holds 108,000 images. The discs can also store movies; this is accomplished by storing individual frames that are played back at the rate of 30 frames per second. The images can be recorded in color or black and white and are stored in *analog* format. This means that the images from videodisc cannot be shown on most computer monitors without first being converted to digital form.

Developers of videodiscs must first assemble the slides, video clips, and movies to be put on the disc. The materials must then be sent to a high-tech processing lab for mastering. The cost of mastering is determined, in part, by the amount of information stored but usually is in the range of tens to hundreds of thousands of dollars. Once the master is created, copies can be made for $20 or less. (When you buy a videodisc, however, you pay more than $20 because the producer is recovering the development and mastering costs.)

To retrieve the information, you need a videodisc player ($500–$1,000) and a color monitor or TV. Most museums and classrooms have videodisc players and are already familiar with the use of videodisc images. Gómez-Pompa and Plummer (1993) have successfully employed videodisc technology for floristic work and point out that the technology is accessible to even the most impoverished countries for both production and viewing.

CD-ROM

A newer storage medium is CD-ROM, the acronym for *c*ompact *disc-read only memory*. These are similar in size and technology to audio CDs. They can store text, graphics, audio, and visual images. They can also store computer programs. CDs have enormous storage capacity (500–700 megabytes) when compared to a floppy disk (0.7 to 2.4

megabytes). The number of images that can be stored depends on the size and resolution of the images: small images at low resolution require less memory than large images at high resolution.

Images are stored in *digital format*. This means that the images can be imported directly into a computer and even exported millions of miles by Internet. Because each image, especially a high-resolution image needed for research, requires a huge amount of digital information, transmission of an image takes up equally large amounts of "bandwidth" during transmission. This causes delays in transmission that are nearly intolerable at this time. As compression algorithms and connections to Internet are improved, transmission of high-resolution images will be more practical. With the images in digital format, scientists and teachers using the proper software can crop, enhance, manipulate, and insert the images into word-processing documents and interactive lessons and publications.

Movies can be stored on CD-ROMs, too, but currently the playback of movies is relatively slow and "jerky," reminiscent of the Charlie Chaplin era of motion pictures. This is because natural motion requires the display of individual frames at a minimum of 30 frames per second. Existing compression-decompression algorithms, coupled with the speed of some CD players and computers, do not allow the display of frames at this rate. Within a short time (probably months by the time this is published) we have been promised that compression and decompression of the digital images will be much faster and more satisfactory.

Resolution of digitized images at the present time is not as good as with videodisc images. It is marginally acceptable for scientific purposes but satisfactory for most educational purposes.

CD-ROM information is retrieved by computers and displayed on monitors. Most classrooms, libraries, and museums have computers and many have CD-ROMs but few have the computer memory capacity and processing speed required for the number of images that are needed for most applications.

Until only a few months ago we would have said that CD-ROMs, like videodiscs, can be made only in specialized laboratories, but now most universities have the technology to make CD-ROMs. The cost of these "one-offs" is only about $20, not including the development costs or capital investment in hardware. Once the master CD is prepared, multiple copies are very inexpensive, approximately $1 to $2 each. CD-ROMs can be read but not altered by the user. This is at once a bane and blessing for the kind of database we are discussing: if the information is correct, we probably would not want most users to alter it; on the other hand, if there are errors, these cannot be corrected. It is important to recognize, however, that we have the same concern with errors that creep into print media.

In summary, there are advantages and disadvantages to the newer visual image archival technologies currently available. In planning for the future, however, it is safe to assume that digital images (CD-ROM) are the wave of the future because they can be stored, disseminated, and retrieved by computers.

Development of Visual Archive Projects

Although we can safely assume that stepped-up efforts to catalog and study biodiversity, nationally and globally, will utilize the more traditional documentation techniques, it seems prudent to test the efficacy of using videodisc or CD-ROM technology or both for archiving visual images on a much smaller scale. This testing should begin immediately; we should not waste time waiting for the technology to improve. The rest of this chapter outlines the management of such projects.

Much planning is required to assure that large-scale projects are well designed *before* they are launched. For example, someone recently suggested that there should be a voucher herbarium specimen for every photo placed on a CD-ROM or videodisc. This absolutely must be known in advance because if we were to begin by collecting slides and *then* decided that vouchers are necessary, there would be no way of collecting the vouchers (except for the tree photographed in your backyard!) and the slide would be rendered useless. Many other aspects of a project require the same kind of careful preplanning and, although we may not be able to plan for every eventuality, we ought to plan as well as possible.

We have just begun to discuss such a project in Ohio under the leadership of the Ohio Biological Survey and The Ohio State University Herbarium. The ultimate goal is to archive images of the flora of Ohio. Limiting it to one midwestern state already delimits the scope, but my guess is that we will eventually identify a pilot project within that larger goal, perhaps limiting our archive to images of the trees of Ohio or the fungi of Ohio or some other taxonomic group. Of course the project could be cut into pieces by entirely different criteria: the prairie flora of Ohio, flora of bogs and fens, wetland flora, aquatic plants of Lake Erie, and so on. Any of these pilot projects would serve the purpose of "getting the kinks out" of a new and complex technology applied to a new purpose.

There are many facets to the process—from planning to production to distribution. The several steps described below should be sufficient for a project on any scale.

Assign Leadership Responsibility

Leadership of the project must be assigned to those with appropriate scientific credentials and proven management skills. This is essential to the success of the project in scientific terms but will also be a major consideration of funding agencies. Computer experts have much to contribute to such a project but, in my judgment, if they do not also have expertise in plant systematics, they will have difficulty gaining the cooperation and confidence of biologists. To have scientific credibility and reliability, the project must be directed by subject-matter specialists.

Management skills are no less important. Such projects require an enormous amount of cooperation and coordination from a very diverse group of specialists. Leaders will need access to experts in fund-raising and fund management, technical resources (computers, photographic equipment and processing, herbarium collections, communication systems, graphic artists, and so on), as well as all the usual purchasing agents, accountants, and human services professionals that support any research project. These requirements could be met in most major universities but might also be found in larger museums, libraries, or botanical gardens. A statewide project might logically be managed by one or more of the state's research universities in cooperation with the state's biological survey.

Appoint an Advisory Committee

As noted above, the project managers must rely on a large retinue of biologists, educators, and technicians. Representatives of these experts should be appointed to an advisory committee that will be responsible for planning the project and bringing it to fruition. Plant taxonomists, ecologists, morphologists, experts in major taxa, scientist/teacher users (schools, universities, museums, state agencies), and those with technical expertise must be brought into the project early on as full partners. Most projects would be able to identify special groups that should be represented on the advisory committee. For the Ohio Flora Archive Project this probably would include representatives of the Ohio Department of Natural Resources, Nature Conservancy, Ohio Native Plant Society, Ohio Academy of Sciences, and others.

Establish Goals

One of the first steps of the advisory committee would be to define the scope of the project in terms of its goals—biological, pedagogical, and technological.

Biological goals   What do research biologists expect to gain from the project? Decisions the advisory committee must make include the following:

1. the scope of the project in terms of *taxonomic groups* (vascular? nonvascular? fungi? algae? bacteria?), *habit* (trees and shrubs only? wildflowers only?), *geography* (Ohio? unglaciated Ohio? northeastern United States?, etc.), *ecosystem* (deciduous forest? prairie? wetland?, etc.);
2. number and kinds of images of each species (e.g., habitat, habit, flower, fruit, seed, dissections, anatomical detail, morphological detail, summer versus winter conditions, pathologies, etc.);
3. the need for voucher specimens of the images; and
4. the need for distribution maps.

Pedagogical goals   How will the archive be used for teaching? Thinking always of both formal and informal educational settings, the advisory committee must decide on the following:

1. level of sophistication of the learners (professional biologists only? college students? high school students? elementary students? general public? all of the above?);
2. level of sophistication of the teachers (experience/comfort with the technology? experience/comfort with the subject matter? need for glossary? need for training?);
3. kinds of courses to be supported (general science? general biology? plant taxonomy? plant morphology? ecology?);

4. kinds of uses (to support lectures/presentations? to support interactive lessons/museum displays? to support interactive research? still frames only? movies, too?)
5. software (software that will be needed, if any is needed, to make the archive suitable for instruction; will this software be included in the archive, be provided in a separate format, or be provided by the user? will it be developed as part of the project or by others?); and
6. copyright and permission (copyright or not? licensing agreements? permission for use under what circumstances?).

Technological goals   What technologies best support the biological and pedagogical goals? In considering these, the advisory committee must keep in mind the cost/benefit ratios of various technical alternatives, not only the cost of project development but also the cost to the ultimate users. Some of the technical considerations include:

1. digital versus analog format (i.e., CD-ROM or other digital storage technology versus videodisc);
2. size of image display and resolution (related issues);
3. stills versus movies (quality of movies, if they are needed for the educational goals, varies with the technology but may be satisfactory in all formats by the time the project is ready to be launched);
4. color quality (various color standards need to be considered);
5. graphics format (there are many standards/formats for graphic images; one needs to be chosen);
6. user accessibility (locked? unlocked? unlocked at what level?);
7. many technological decisions related to the production phase of the project; and
8. method of mastering and duplication (choice of format may or may not require different methods and costs for mastering and duplication).

Estimate Costs of the Project and Apply for Funding

Cost estimates depend mainly on goals defined above but there are some additional considerations: (1) whether the final product should be marketed/distributed directly or through a publisher; (2) whether "free" copies will be distributed to designated institutions; (3) the estimated number of copies needed; and (4) the cost of revisions, new editions, and so on.

Most projects will have an assortment of funding options: (1) total funding from one source; (2) funding from multiple sources (perhaps divided by region, subject matter, technology, ultimate use, etc.); and (3) venture capital from grants or loans (with cost recovery from sales, licensing fees, etc.).

Establish Deadlines

The archive could be published in "phases" or postponed until a full set of images has been assembled. Many similar projects decide to publish whatever is gleaned from contributors in the first year, complete or not. Each year new images are added to the original collection (the most cost-effective way is to add them to the end of the previous collection), which is then published as a new edition. These new technologies will allow that flexibility because the images can be accessed randomly so there is no absolute need to organize and reorganize the images when new images are added (though many would argue that the archive is much easier to use when there is some organization to it).

Gather and Select the Visual Images

Certain policies and procedures must be in place before the first image is accepted. For example, if the images are to be given to the project with full copyright release, a permission form is required of the contributors (there are numerous examples of such forms from other projects). Of course some method of cataloging and storing the images themselves is needed, as well as a way of storing information about the images. Difficult decisions about which images to include and which to reject will arise. [One example of an interesting question that has already come up: Should the Ohio Flora Archive include a photo of a species taken in West Virginia if that species also grows in Ohio?] If it is decided that voucher specimens must accompany the photos, where will the vouchers be stored and how will they be tied to the photos?

Process the Images

There are (expensive!) cameras on the market that record and store the images in digital form rather than on film, but one can safely assume that most images will come from new or existing slides. The slides or prints will need to be converted to a form suitable for the medium chosen. For example, if CD-ROM is selected, the images must be digitized. This is a very time-consuming task and is also expensive. The advisory committee needs to decide early on whether this will be done by the project or contracted to a vendor. If done by the project, the costs of digitizing equipment and staff need to be included in the project's budget. If done by a vendor, the vendor's charges need to be in the budget.

Create the Master Archive

Again, how this is done depends on the medium and also on the goals of the project. The advisory committee needs to consider whether this can be done "in house" or through a vendor. Either way, this is another budget item.

Press and Distribute Copies

Preproduction testing of the final product is very important but not always easy depending on the technology adopted. If CD-ROM is chosen, it is relatively easy to make "one-offs," single copies of the CD-ROM (approximately $20 each) that could be used for field-testing before large quantities are duplicated. If videodisc is chosen, it is very difficult to field-test before copies are pressed. If errors are found after pressing, it is impossible to correct them short of mastering a new videodisc.

An important consideration about distribution is this: Be sure to keep accurate address and telephone records of all recipients of the archive. This is important because (1) you will want to make updates available to them; (2) you may need to notify them of errors; (3) you may want to invite their participation in a workshop (e.g., at the annual meeting of the state science academy) where users could share their innovative uses of the archive with other biologists/teachers; and (4) you may need to conduct a mail or phone survey for feedback about the way the archive is used, its value in the classroom, and so on. If you need funds to update or expand the project, the granting agency will almost certainly ask for proof of the archive's value to users.

Update the Archive

The planning phase of the project is not too early to discuss the question of updating the archive or expanding it. Many of the steps listed above bear some relationship to the method and frequency of issuing the archive in new editions.

Obviously, these procedures will need to be modified for a specific project but should help the managers avoid some of the pitfalls and assure that the visual archive will make a valuable contribution to research and education.

Literature Cited

Anonymous. 1994. *Systematics Agenda 2000: Charting the Biosphere. Technical Report.* New York: Systematics Agenda 2000.

Gómez-Pompa, A. and O. E. Plummer. 1993. A view of the future for floristic research. In F. A. Bisby, G. F. Russell, and R. J. Pankhurst, eds., *Designs for a Global Plant Species Information System*, Systematics Assoc. special vol. 48, pp. 83–93. Oxford: Clarendon Press.

Kim, K. C. and L. Knutson, eds. 1986. *Foundations for a National Biological Survey.* Lawrence, Kans.: Association of Systematics Collections.

Lucas, G. Ll. 1993. The need for a worldwide botanical reference system. In Bisby, Russell, and Pankhurst, *Designs*, pp. 9–12.

Soulé, Michael E. 1986. Conservation biology and the "real world." In M. Soulé, ed., *Conservation Biology: The Science of Scarcity and Diversity*, pp. 1–11. Sunderland, Mass.: Sinauer.

Stern, W. L. and K. L. Chambers. 1960. The citation of wood specimens and herbarium vouchers in anatomical research. *Taxon* 9:7–13.

Unasylva. 1992. Tenth World Forestry Congress—Dossier. 43 (1): 3–9.

U.S. Congress, Office of Technology Assessment. 1992. *Combined Summaries: Technologies to Sustain Tropical Forest Resources and Biological Diversity*. OTA-F-515. Washington, D.C.: U.S. Government Printing Office.

Wilson, E. O. 1992. *The Diversity of Life.* New York: Norton.

Wilson, E. O. and F. M. Peter, eds. 1988. *Biodiversity*. Washington, D.C.: National Academy Press.

8

Implications of Virtual Technologies for Cognitive Diversity

DON STREDNEY

The need for continued documentation of species diversity is evident. As we turn to synthetic methods to represent and disseminate collected information, the technologies of virtual simulation become more integral to the repertoire of the herbarium, the research lab, and the classroom. Algorithms capable of generating the structural fidelity required are being developed. These programs will provide insight into plant development and diversification.

The possibilities for elucidating how the brain encodes and decodes sensory information are certainly expanding. As we broaden this understanding, we increase our ability to create synthetic representations of objects, phenomena, and environments. Our abilities to create computer-generated virtual models of the world and to perform operations to these models has already led to new areas of scientific, artistic, and educational inquiry (Cooperrider and Cooperrider 1994). The ability to generate virtual simulations and representations with increasing accuracy and precision (Oppenheimer 1986; Fowler, Prusinkiewicz, and Batljes 1992; Prusinkiewicz, James, and Mech 1994) may prove to be one of the most significant achievements of this century.

This paper presents an interdisciplinary overview of the issues of representation, specifically their use in the generation of virtual representations. The implications of these issues to interface design, and the significance of these designs to emerging technologies and their use in learning, will be discussed.

Definitions of Reality

Much of the difficulty in defining virtual reality lies in defining the term *reality*. Trying to answer such questions as "What is real?" or "What is reality?" leaves us caught in a morass of ontological philosophy.

Moreover, in order to contemplate reality we must grapple with another enigmatic concept, consciousness. Is consciousness equivalent to awareness? Much of the information encoded, processed, and decoded by the brain is unconscious. Even today, in all our modernity, our answers to questions of consciousness and reality are those of our ancestors. Inevitably we find that definitions of consciousness and reality derive from our unique perspectives, based on our specific experiences and genetic makeup. We come to understand that reality and consciousness are constructs of the mind.

Problems in defining virtual reality are exacerbated with technological considerations. For instance, is it virtual reality if it is not immersive? In other words, should the technology completely surround the senses in order to be considered immersive? We assume that if it surrounds the senses, it is more likely to be engrossing. Should virtual reality involve all the senses, including tactile and olfactory? Perhaps these prerequisites are too confining. Just as various levels of reality exist, there are various levels of virtual reality.

Ultimately, the value of virtual reality or representations is what the user gains from the experience. Whether we are engrossed or absorbed in the content depends on the context created and, to a great extent, our own involvement. Complete absorption may or may not require immersion of the senses. Anyone who has been moved by the written word, a theatrical presentation, or a dream, or has witnessed a tragic scene, realizes that immersion of all the senses is not a prerequisite for a meaningful experience, one in which the individual is completely and consciously absorbed.

Perception and Consciousness

Is reality the limited stream of cognition we have termed *consciousness*? Recent developments in the fields of perception and neuroscience have shown that there are various levels of consciousness. The brain is anatomically and functionally adaptable to a degree not previously considered (Grobstein and Chow 1989; Crick and Koch 1992; Kandel and Hawkins 1992). In an attempt to understand this plasticity, we find that much of our perception occurs through unconscious processes (Gazzamiga and LeDoux 1978). Wolfe has demonstrated that the visual system comprises a set of subsystems and that many of these processes are in fact hidden: "No amount of introspection can make us aware of the subsystems themselves" (Wolfe 1983:94). Through these findings we conclude that our perceptions, like our concepts of reality, are only a thin slice of the world around us. Schopenhauer (in Durant 1960:312) previously suggested that "consciousness is the mere surface of our minds, of which, as the earth, we do not know the inside, but only the crust."

The brain is a wonderful result of evolution, an organ that embodies several complex systems whose basic tasks are to analyze the information it senses and to coordinate and control body functions and movements. Our brains take in a limited slice of the available world, separate signal from noise, analyze and assign some meaning and significance to the information, and reconstruct a plausible, workable model of the world. The brain constructs a paradigm of the world around us, as opposed to simply reflecting the "outside" world brought in by our senses. It then compares and adjusts its model in order to effectively move within the perceived world as well as to change it. Throughout one's existence, the model is checked, rechecked, and modified as the brain seeks congruency and consonance. The world appears "real" not because it actually looks or is that way, but because our brains construct it that way. Brains of animals "see" different aspects of the world. It has been demonstrated that brains change the way their sense organs can sample the world (Barlow 1990).

Through comparative studies of animal models, transitional changes in developing systems, and fossil records, we realize that natural selection has and continues to play a major role in the development of perception and thus the various views of reality held by living creatures. Other species sample other bandwidths from the spectrum of life. Out of different environments evolve different sensory systems with acuity and focus adjusted to assure survival in that particular environment. From the various paradigms created by the brains of living creatures originate different expressions of that paradigm.

Mental Schemes and Representation

The brain processes information through what is known as hierarchical feature processing (Hughes and Sprague 1989). The brain extracts information such as edges, colors, and movements and processes them separately (Hubel and Weisel 1979). Further up the dendritic tree, the information is analyzed and consolidated (Livingston 1988; Livingston and Hubel 1988). Information is then linked to spatial and temporal, behavioral and environmental contexts (Grobstein and Chow 1989). Some have suggested that the integration of sense information from two disparate sources, such as the use of sight and balance to perceive motion, provides us with a unique sense of unification (Henn, Cohen, and Young 1980). However, the mechanism of unifying information and the putative site of integration remains elusive (Wilson, O'Scalaidhe, and Goldman-Rakic 1993).

An intriguing neuroanatomical model has been proposed to explain such unique behaviors as the perplexing phenomenon of phantom pains (Melzack 1992). Melzack postulates the existence of a genetically programmed cortical network dedicated to information concerning the body. This network is manifested as the concept of self. Individuals who have never had limbs, such as thalidomide victims, often internally visualize their arms swinging as they walk. Pathologies to the system are manifested clinically in individuals who do not recognize their own limbs. They are distraught that someone else's limbs are in bed with them and continually attempt to throw the unknown limb out of their bed. Amputees, tortured by the perceived itching of the limbs they have lost, actually can benefit from internally visualizing themselves scratching the affected limb. Others cite similar evidence of the value of previsualizing one's body moving through the motions as a method to learn certain athletic activities (Kosslyn 1983; Finke 1986). This previsualization may prime the cortical network proposed by Melzack, the neural system that gives rise to the concept of body and the manifestation of self, and provide an unconscious form of learning.

Whether there is an integrated site processing final representations of information within the brain remains to be seen. It appears doubtful that a single site of integrated information will be found. The hypothesized area may be only a concept, a construct of the mind itself. Perhaps it is this sense of unification as proposed by Henn and others (Henn, Cohen, and Young 1980) that gives rise to the concept of central integration.

As neuroscience continues to search for the mechanisms of perception and consciousness, memory consolidation and the actual site of memory, we gain valuable insight into how the brain encodes, stores, and decodes information. This knowledge will be extremely applicable to future virtual technologies. In order to effectively create virtual representations, we must not only know how to represent the world through mathematical descriptions and photorealistic renditions, but also know how to identify and mimic the schemes the brain uses in processing information. Our goal should not be to re-create the world or objective reality, which is an intractable problem computationally, but rather to communicate meaningful information from one brain to another. Through the use of mental and artistic schema, we can synthesize a plausible and tractable representation of reality.

Animal Models and the Evolution of Representation

Just as comparative studies have been useful to us for constructing a model of brain processing, so too are they helpful in elucidating the methods of representation. Animals have evolved elaborate mechanisms of representation to communicate information to one another. Protective mimicry to deceive and flash coloration to confuse predators, as well as cryptic coloration to conceal and camouflage, are widely employed in the animal world.

Innate release mechanisms have evolved to process specific environmental stimuli in order to produce effective behavioral responses, such as the avoidance response or mating rituals (Lorenz 1950, 1965; Tinbergen 1951). These sign stimulus and fixed action patterns are neural pathways that respond only to specific cues observed in the environment. They present us with models by which the brain can store information and subsequent behaviors that are rapidly decoded on the presentation of a specific stimulus or certain stimuli within a specific context, such as spatial information (pattern and form), light frequency (color), and temporal sequence (motion). A similar theory has been proposed to explain phenomena of rapid change such as avalanches and information propagation through the brain (Bak and Chen 1991). This theory, known as self-organized criticality, reflects the brain's parsimonious management of information. As the brain is provided with a specific cue, a large amount of stored information comes rapidly to awareness. Such a phenomenon occurs with smells, as when the smell of apple pie takes us back to our grandmothers' kitchens with their checkered tablecloths and other sensory accoutrements.

This use of association is an extremely powerful technique for storing and retrieving information. In humans, infants around five months of age will anticipate an entire object when only shown a part. This ability to attribute whole information from only a part is the basis of symbolic representation and is known as metonymy (White 1989).

Randal White provides a provocative view of the emergence of symbolic representation in humans. The archaeological records of human existence on earth, which date back 2.5 million years, contain predominantly utilitarian objects (Knect, Pike-Tay, and White 1993). It was not until around thirty-five thousand years ago, in what is known as the Aurignacian Period, that we find a sudden appearance of objects that have an intentional decorative or aesthetic function. White postulates that the increase in symbolic forms resulted from cultural processes. As societal status developed, so did the techniques for metonymy. White (1989:98) states that "close to the heart of these developments is an increased ability to think in—and communicate by means of—special visual images." He proposes that this increased competence might help to explain the burst of technological progress that characterized the Aurignacian period. White surmises that this increase cannot be attributed to biological or neurological processes. Modern man, *Homo sapiens*, evolved some hundred thousand years ago with definite neurological advantages, primarily cranial capacity, over, say, *Homo neanderthalensis*, or Neanderthal man, whose skeletal remains have been found in Western Europe dating back some eighty thousand years. However, skeletal remains that delineate cranial capacity clearly cannot account for neurological developments such as general plasticity and the establishment of new associative pathways, which do not survive intact. Animals had been making use of associations for millions of years. Perhaps our ancestors evolved new associations that built on sign stimuli and innate releasing mechanisms and allowed them to formulate representations that not only mimic nature but also *decode* and *express* internal abstract representations in new ways, specifically through the manipulation of form and pictorial representation.

Brooke Hindle, a science historian who studies the Industrial Revolution, believes that the advances made during that time resulted from the significance of the techniques for image manipulation, such as the use of exploded drawings for conveying technical innovations and the development of photography (White 1989). A similar claim can be made of the Renaissance and the development and subsequent use of perspective in pictorial representation. If so, the power to electronically synthesize virtual representations presents a quantum leap in our methods to communicate. The television age has merely been a prelude to the expanded use of synthetic sensory information.

Artistic Metaphors of Mental Representations

Complex schema and visual metaphors have been used for millennium to represent information. Since the early records of the Venus of Willendorf and the paintings at Lascoux, artists have adopted elaborate schema for expressing information. The variety of schema can be attributed to the depiction of what is *conceived*, rather than what is *perceived*. The diversity of representation can be concluded from individual and cultural influences, at both the genetic and environmental level.

The development of artistic schemes has led Western art from realistic representation to illusion to trompe-l'oeil painting to abstract expressionism. However, the effectiveness of the work always relied on the participation of the viewer, or what Gombrich terms *the beholder's share* (Gombrich 1989). All representations, whether abstract or representational, attempt to communicate. Each elicits the release of internally held information, the beholder's share. Whether modeling a two-dimensional surface to represent a three-dimensional object or space, or the use of expressive gestures or complex symbolism, the artist uses cues of various representational levels to tap into a vast reservoir of stored information.

Through the use of artistic representation over the centuries, we have developed a sophisticated repertoire for creating and understanding visual information. With the advent of motion pictures in the late nineteenth century, we introduced an inherent method to depict temporal information. The brilliant cinematic effects introduced by D. W. Griffith, including the close-up, cut, and parallel editing (which he attributes to Dickens), have become commonplace in our everyday visual experience (Pudovkin 1933). Griffith freed us from the running narrative and provided us with visual "chunks" that permitted a more engaging method to depict a story. These cinematic techniques mimic the brain's methods of leaping from significance to significance, passing over the noise with unconscious processing. Griffith's techniques made his films appear, according to his audiences, more lifelike and more engaging. These methods may be more engaging not only because they reflect the brain's methods of processing information but because the brain must fill in missing information between presented significance. The associations are not completely apparent but must entail further involvement by the viewer, albeit unconsciously.

It is intriguing to consider the similarity between artistic methods and methods employed by the brain to encode and process information. Have we mimicked the brain's methods of processing visual information in the schemes we use to create and express visual information? Does the brain reveal its mechanisms of information storage in its methods of expressing information? Hubel and Weisel (Weisel, Hubel, and Lam 1974) discovered that Van Gogh intuitively exploited the methods of information processing in the visual cortex. Griffiths's use of nonserial imagery is similar to imagery experienced in dreaming. Finke has demonstrated that methods to create internal visual imagery employ the same brain mechanisms used in perceiving visual information (Finke 1986).

We must also consider that perception and representation are reflexive. The brain constructs the world around it and is influenced during its development by what it perceives. Kosslyn (1989) has defined an image-percept analog, which theorizes that mental representations assimilate constraints learned in the physical world, such as time, scale, and distance. However, in an electronically synthesized world, as in a dream, these variables can be modified. Today's generation has an unprecedented visual and temporal sophistication. As we use more sophisticated visuals in our electronic media, our children have and will create and synthesize increasingly sophisticated models of their world (e.g., Horgan 1993). Further study is needed to understand the power of representations and their influence on the development of an individual's world model.

The pursuit of verisimilitude in computer-generated visuals has a unique parallel to the historic pursuit of realistic representation in art.

Over the short history of computer graphics, the main research pursuit has been to generate photorealistic imagery. Currently, rendering methods are far in advance of methods for generating and manipulating data. Present renderers are capable of creating stunning examples of photorealistic imagery. This technology to synthesize, modify, and disseminate digital visual information allows us to model our world with unprecedented precision and accuracy. Digital methods provide seamless merging of visual information, so obscure that a great deal of the computer graphics used in today's television advertising and cinematic special effects go unperceived as such.

For many, "seeing is believing." Yet this ability to create realistic representations offers a dangerous opportunity for deceit (Mitchell 1994). Inaccuracies in synthesized imagery can be especially damaging in the fields of forensic, medical, and scientific visualization. Extraordinary diligence will be necessary to detect the misuse of the technology to prejudice and persuade. Juries are shown reconstructions of crime through computer simulation. Just as in cinema and television, how the information is represented and the context in which it is presented can be extremely prejudicial. The solution to misuse of the technology will be dissemination of data used to create the imagery or environment. Scientists will need to corroborate the findings by using the same data. The defense and the prosecution will need to share digital data sets used in court.

Much research has been devoted to exploring the levels of representations for facilitating learning, most specifically computer-generated imagery (Wagner, Fewanda, and Greenberg 1992). Foley (1988) has demonstrated that there is a limit to the perceptual cues that an individual needs to do certain tasks. In his study, faceted shaded objects allow the user to match a structure with its rotated counterpart, by an increase of 20 percent over hidden line or vector graphics. Smooth shaded objects, however, offered no improvement for recognizing rotated objects.

Lintern (1992) has shown that simple graphics may be more helpful to novice flyers when using simulations, and that they may actually reduce practice time by 15 percent. Lintern suggests that overly realistic displays may be too distracting for users and sometimes overwhelm them, causing them to panic. When, however, does increased realism assist in conveying a more accurate depiction? Eventually one would want to train on a system that provides enough realism in the simulated environment to maximize transfer to the "real" world. Experts subtly and intuitively use nuances in the environment to effectively assess and interact with their environment. At one time consciously sought out, these cues are now unconsciously processed.

Learning

One phenomenal aspect of human cognition is the ability of children to learn language. Thompson (1989) estimates that from the age of one through fifteen, a child learns on *average* two hundred words a day. How do children process such large amounts of information with no apparent rule base? Nondeliberate learning is known as incidental learning. As children play, their brains continuously process information. Whether it is the titration of fine motor coordination or the subtle understanding of how different materials work and interact, the brain is learning.

Similar unconscious processing can be seen in selective visual attention. Visual attention occurs in two stages (Koch and Ullman 1985). The first stage is called *preattentive*. It is a very fast processing of the entire visual field that occurs in parallel. Such attention determines colors, depth, motion direction, and local orientations. The second stage is called *attentive*, or *focus of attention*. It is slower and occurs serially, focusing on particular regions of the visual field. In our educational methods, however, we have adopted this attentive stage as the main modality for learning, whereas the preattentive stage has been neglected. It is the preattentive stage, the rapid and parallel method for processing information, that experts may be using in evaluating situations. This unconscious process may be rapidly evaluating subtle nuances in the environment and flagging the conscious process for any abnormalities, novelties, or deviances from stored patterns.

Representation has been defined as either active or latent (Crick and Koch 1992). An active representation is that which is perceived, one that may employ working memory. Latent representation is based on stored information. It employs referenced memory. As we compare active representations (percepts) to latent representations (concepts), context becomes extremely important. Deadwyler (1989) has delineated the mechanism involving an area of the brain, the hippocampus, which processes information in relation to environmental or behavioral context. The context in which something is viewed affects the way it is perceived. McGreevy (1992) states that context "is the fabric in which a thread lies." In addition, the cognitive expectations of the observer, the beholder's share, strongly influence perceptions (Wolfe 1988). By creating more sophisticated imagery, we may facilitate the brain's ability to assimilate novel information by introducing cues that establish links to internally held contexts. Imagery that includes detailed spatial and color cues, as well as motion, convey information that the brain expects to see. These realistic cues facilitate the brain to encode information rapidly and may provide the difference between learning and experience. They aid the individual in participating with content, albeit only conceptually and on an unconscious level.

Cognitive Diversity

Within a certain domain, most of our perceptions are similar. Beyond that range, however, our perceptions are singular, defined by the unique experiences that each of us undergoes. What is required in emerging virtual systems are interactive interfaces that permit the individual to titrate the level of complexity in the visual cues provided in the simulation and to modulate the speed at which the information is presented and explored. Studies in neuroscience have demonstrated physiological mechanisms for individuality (Kandel 1985; Kandel and Hawkins 1992). We should capitalize on the flexibility of the computer medium and design interfaces that allow users to augment their unique mental capacities. The user should be able to create the level of realism that provides the visual schema best suited to their understanding, facilitates their assimilation of new information, and aids in the transfer of that knowledge to a practical setting.

If we standardize the methods by which we interface with virtual systems, we are doing nothing more than the current practice of the proscenium verbal presentation of information commonly used in education. This method is biased for those who learn in a serial fashion. However, it completely leaves out individuals who specialize in processing information nonserially and in parallel.

As reality is a construct of the brain, the shared reality we experience is the result of equivalencies in genetic makeup. In our shared perceptions, the way we see the world represents the brain's way of creating representations of our existence that are meaningful. By creating imagery that mimics the way we see, we may augment the assimilation of new information. By developing referents that build on natural representations, we may be facilitating the sharing of information across disciplines. We present visual cues that the brain *expects* to see.

However, genetic makeup is congruent only to a degree, and as experiences are unique, so too are the methods by which we perceive. Discoveries in neuroscience have provided a model for the biological expression of individuality. We should respect this diversity and nurture it, for the ability to see the world in new ways provides us with an instrument of change, a mechanism for both collective and individual growth. Emerging developments in virtual systems provide a method by which to communicate information effectively and respect and nurture the cognitive diversity that is the hallmark of human achievement.

References

Bak, P. and K. Chen. 1991. Self-organized criticality. *Sci. Am.* 264 (1): 46–53.

Barlow, R. B. 1990. What the brain tells the eye. *Sci. Am.* 262 (4): 90–95.

Cooperrider, M. K. and T. S. Cooperrider. 1994. History and computerization of the Kent State University Herbarium. *Ohio J. Sci.* 94 (1): 24–28.

Crick, F. and C. Koch. 1992. The problem of consciousness. *Sci. Am.* 267 (3): 152–159.

Deadwyler, S. A. 1989. Evoked potentials in the hippocampus and learning. In *Learning and Memory–Rdgs. Encycl. Neurosci.*, 54–55. Boston: Birkhauser.

Durant, W. 1960. *The Story of Philosophy*. New York: Washington Square Press.

Finke, R. A. 1986. Mental imagery and the visual system. In *The Perceptual World–Rdgs. Sci. Am.*, 179–189. New York: Freeman.

Foley, J. D. 1988. Interfaces for advanced computing. *Trends in Computing* 1:62–67.

Fowler, D. R., P. Prusinkiewicz, and J. Batljes. 1992. A collision-based model of spiral phyllotaxis. *Proc. Comp. Grphcs.* 26: 361–368.

Gazzamiga, M. S. and J. E. LeDoux. 1978. *The Integrated Mind*. New York: Plenum.

Gombrich, E. H. 1989. *Art and Illusion: A Study in the Psychology of Pictorial Representation*. Princeton, N.J.: Princeton University Press.

Grobstein, P. and K. L. Chow. 1989. Visual system development and plasticity. In *Learning and Memory–Rdgs. Encycl. Neurosci.*, 56–58. Boston: Birkhauser.

Henn V., B. Cohen, and L. R. Young. 1980. Visual-vestibular interaction in motion perception and the generation of nystagmus. *Neuroscience Res. Prog. Bull.* 18:457–651.

Horgan, J. 1993. The Death of the Proof. *Sci. Am.* 269 (4): 92–103.

Hubel, D. H. and T. N. Wiesel. 1979. Brain mechanisms of vision. *Sci. Am.* 241(3): 150–162.

Hughes, H. C. and J. M. Sprague. 1989. Visual learning, pattern, and form perception: Central mechanism. In *Learning and Memory–Rdgs. Encycl. Neurosci.*, 59–61. Boston: Birkhauser.

Kandel, E. R. 1985. Cellular mechanisms of learning and the biological basis of individuality. In E. R. Kandel and J. H. Schwartz, eds., *Principles of Neuroscience*, 2d ed., 816–833. New York: Elsevier.

Kandel, E. and R. D. Hawkins. 1992. The biological basis of learning and individuality. *Sci. Am.* 267 (3): 78–86.

Knect, H., A. Pike-Tay, and R. White, eds. 1993. *Before Lascaux: The Complex Records of the Early Upper Paleolithic*. Boca Raton, Fla.: CRC Press.

Koch, C. and S. Ullman. 1985. Shifts in selective visual attention: Towards the underlying neural circuitry. *Human Neurobiol.* 4:219–227.

Kosslyn, S. M. 1983. *Ghosts in the Mind's Machine*. New York: Norton.

———. 1989. Mental imagery. In *Learning and Memory–Rdgs. Encycl. Neurosci.*, 75–76. Boston: Birkhauser.

Lintern. 1992. Back-to-basics training. *Tech. Rev.* 80 (October).

Livingston, M. S. 1988. Art, illusion, and the visual system. *Sci. Am.* 258 (1): 78–85.

Livingston, M. and D. Hubel. 1988. Segregation of form, color, movement, and depth: Anatomy, physiology, and perception. *Science* 240:740–749.

Lorenz, K. Z. 1950. The comparative method of studying innate behavior patterns. *Symp. Soc. Exp. Biol.* 4: 221–268.

———. 1965. *Evolution and Modification of Behavior*. Chicago: University of Chicago Press.

McGreevy, M. W. 1992. The presence of field geologist in mars-like terrain. *Presence–Teleop. Virtual Envir.* 1 (4): 375–403.

Melzack, R. 1992. Phantom limbs. *Sci. Am.* 266 (4): 120–126.

Mitchell, W. J. 1994. When is seeing believing? *Sci. Am.* 270 (2): 68–73.

Oppenheimer, P. E. 1986. Real-time design and animation of fractal plants and trees. *Proc. Comp. Grphcs.* 20: 55–64.

Prusinkiewicz P., M. James, and R. Mech. 1994. Synthetic topiary. *Proc. Comp. Grphcs.* 28: 351–358.

Pudovkin, V. I. 1933. *Film Technique*. London: Newes.

Thompson, R. F. 1989. Introduction. In *Learning and Memory–Rdgs. Encycl. Neurosci.* Boston: Birkhauser.

Tinbergen, N. 1951. *The Study of Instinct*. Oxford: Clarendon.

Wagner, L. R., J. A. Fewanda, and D. P. Greenberg. 1992. Perceiving spatial relationships in computer-generated images. *IEEE Comp. Grphcs. Apps.* 12 (3): 44–58.

White, R. 1989. Visual thinking in the Ice Age. *Sci. Am.* 261 (1): 92–99.

Weisel, T. N., D. H. Hubel, and D. M. K. Lam. 1974. Autoradiographic demonstration of ocular-dominance columns in the monkey striate cortex by means of transneuronal transport. *Brain Res.* 79:273–279.

Wilson, F.A.W., S. P. O'Scalaidhe, and P. S. Goldman-Rakic. 1993. Dissociation of object and spatial processing domains in primate prefrontal cortex. *Science* 260: 1955–1958.

Wolfe, J. M. 1983. Hidden visual processes. *Sci. Am.* 248 (2): 94–103.

———. 1988. Visual perception: Sensory systems I. In *Visions and Visual Systems–Rdgs. Encycl. Neurosci.*, 93–94. Boston, Birkhauser.

9

Remote Sensing and Geographic Information Systems: Spatial Technologies for Preserving Phytodiversity

MICHAEL N. DEMERS

This chapter examines the potential role of satellite remote sensing and geographic information systems (GIS) technology in the preservation of phytodiversity. Remote sensing is examined primarily from an in situ perspective, that is, in terms of its applicability to evaluate in-place environments on a regional scale. GIS is considered for its applications both in situ and ex situ, as well as for its management, modeling, and integration capabilities. Current utilization of both technologies, although useful, indicates a limited understanding of their potential to investigate, evaluate, and predict future vegetative conditions. This paper illustrates, using hypothetical examples, some simple but powerful techniques available within most commercially available systems, as they might be applied to phytodiversity preservation. It shows the use of remote sensing to provide a synoptic view of the vegetation, often producing maps of higher quality than those produced through time-consuming, ground-survey methods. The use of GIS analysis for preserving phytodiversity is illustrated through

*examples of how visualization, point-in-polygon searches and fre-
quency analysis, map coverage integration (overlay), and buffering
can locate, isolate, and compare important phytosociological data.
Finally, this paper provides some insights into the design considerations for long-term functioning of a national biodiversity GIS.*

Among the more exciting and useful tools available in the conserva-
tionist's toolbox are digital remote sensing and geographic information systems. Both these technologies are capable of very powerful
analysis of regional-scale, spatiotemporal data (Butler, Ludeke, and
Palmer 1990). Each serves a specific purpose—remote sensing provides a monitoring capability, whereas GIS is generally associated
with modeling and analysis. However, both technologies have many
operational similarities, and many software systems offer capabilities
common to both GIS and remote sensing. In fact, perhaps the most
important aspect of both is that they are easily integrated, providing
an exciting superset of capabilities for studying the earth's vegetation
(U.S. Congress 1987).

Remote Sensing for a Synoptic View

Digital remote sensing provides a synoptic view of vegetative characteristics at a single or multiple points in time. Data are usually recorded
as cell-by-cell radiance values for selected portions or bands of the electromagnetic spectrum that act as surrogates for ground-cover condition. This ground-cover condition necessarily reflects the vegetation, its
vigor, phenology, soils, and background, as well as moisture conditions.
Each grid cell, or pixel (which stands for picture element), records a single value for each band of radiation examined. This value is an average
over the entire pixel field of view, often called the instantaneous field of
view (IFOV) (Lillesand and Kiefer 1994). Because each pixel is an average value, the size of the pixel needed for examining specific vegetation
groups or conditions must be chosen carefully.

There is a wide range in the spatial resolution capable of being
sensed with modern satellite remote-sensing devices. Large pixels are
available using the Advanced Very High Resolution Radiometer
(AVHRR) aboard the GOES weather satellite. These large pixels, 1.1
km per side, allow the user to capture information for an entire continent without completely overwhelming the current computers. Finer-

resolution sensors range from the now common LANDSAT multi-spectral scanner data (MSS), with a nominal resolution of 80 meters, to the LANDSAT thematic mapper data (TM), with its 30-meter resolution, down to the 10-meter resolution available from the French Système Photographique pour l'Observation de la Terre (SPOT) satellite. Finer resolution is available using airborne sensing, such as airborne videography and side-looking airborne radar (SLAR) for use with smaller study areas.

A common trade-off in remote sensing is spatial resolution sacrificed to obtain improved temporal resolution. In some instances, large pixel sizes may not be a disadvantage because the larger cell-sized sensor is capable of covering large areas frequently. This improved temporal resolution is probably best exemplified by the GOES AVHRR sensor, which allows sensing as often as four times a day, whereas LANDSAT is limited to a single scene every sixteen days. The improved temporal resolution of AVHRR allows for the development of phenological databases, even permitting improved vegetative classifications by incorporating this phenological information into the classification process. Special arrangements might be made for improved temporal resolutions of special-purpose sensors like airborne videography; however, these special overflights often prove prohibitively expensive, especially if frequent revisiting is necessary.

Because the spectral responses of vegetation types vary greatly, it is often necessary to be highly selective in the portion of the electromagnetic spectrum used for sensing the vegetative condition. The spectral dimensions available through remote sensing are likewise variable and easily obtainable. There are sensors capable of viewing the landscape in wavelengths ranging from thermal and reflected infrared through visible and into the microwave bands. The most commonly applied spectral dimension for vegetative indexes is some combination of the visible and near-infrared bands from LANDSAT and SPOT satellites. Numerous sources of information exist concerning the typical uses of the more common sensor bands aboard the LANDSAT, SPOT, and other satellite systems. I have included as a reference source a current text that has such information readily available (Lillesand and Kiefer 1994).

No matter what data are obtained, or from what satellite, two forms of output are generally available—digital and analog. Hard-copy output is easily inspected visually and requires little training outside of a general knowledge of the area being sensed, its vegetation, and the characteristic responses of the sensor to that type of vegetation. The trend, however, is toward data supplied in digital form so that image-processing programs can manipulate these files to better retrieve the necessary information. These retrieval manipulations range from simple image enhancement techniques, allowing better visual inspection, to supervised and unsupervised classification algorithms that will produce final map output of the necessary categories of data. The classified maps are finally generated as hard-copy output from which decisions can be made.

Ultimately remote sensing offers a large spatial context on which to evaluate vegetative conditions. In addition, this view, if properly created, can act as a surrogate for extensive field data, at least at a regional level. Should individual specimens be needed for classification or for preservation, the classified satellite image provides an excellent framework for creating a stratified sample design methodology. Of course, within limits, vegetational change is obtainable with satellite remote-sensing data.

Some individuals within the remote-sensing community concerned with biodiversity have developed their own research agenda (Stoms and Estes 1993). Concentrating exclusively on species richness as an indicator of biodiversity, these scientists propose that inventory is a necessary first step in evaluating global-change impacts. Their view is that a determination of biophysical controls will assist in modeling efforts and that monitoring and forecasting changes in biophysical factors allow the creation of predictive models of vegetative response to that change.

From a phytodiversity perspective, some of these agenda items are already being implemented through the GAP Analysis programs in the United States and through the EMAP Program of the Environmental Protection Agency (Scott et al. 1990, 1993). The GAP remote-sensing activities, however, are focused on an evaluation of the vegetation as habitat for fauna, predominantly the more visible fauna. The fundamental goals of GAP are to determine gaps in distribution of terrestrial wildlife habitat. This activity is also reflected in other activities of the national biological survey, which include biomonitoring of environmental status and trends (EMAP), Great Lakes fisheries assessment, breeding-bird surveys, and wildlife mortality database (Davis et al. 1990). These activities are not entirely related to remote sensing and will require additional manipulation capabilities beyond digital image processing.

Geographic Information Systems for Data Integration

Remote sensing provides a limited set of data for preserving phytodiversity. Combined with existing and anticipated cataloging and storage techniques, the geographic information system offers an opportunity to integrate both spatial (entity) and associated descriptor (attribute) data. GIS combines its functions of input, editing, archiving, analysis, and output to incorporate data collected either through sensors or through field-related activities. It can add widely varying sources of information such as soils, topography, hydrology, land ownership, and a host of other data sets, input either from maps or collected directly. Perhaps more important, it can, through the use of database technology, tie the actual vegetation to explicit locations on maps, thereby adding the spatial and temporal dimensions to the monitoring of phytodiversity.

Once these data are incorporated into GIS, they can be recalled and viewed singly or as multiple views of each of the data layers (coverages). For example, one could incorporate, on a single computer monitor, the synoptic view of habitat types together with topography and even digital pictures of species taken in the field or as herbarium specimens (fig. 9.1). Additional information concerning the known status of the species and even names and addresses of species specialists for consultations could be incorporated as part of the query output.

Beyond the simple viewing of data known to exist at a given site, most modern GIS allow the scientist to show spot locations of selected taxa, either singly or in combinations. In addition to displaying known locations, tabular data concerning sociability, protection status, known herbarium specimens and locations, and many other pieces of ancillary data concerning the selected taxa can be acquired. Alternatively, it might be necessary to determine what known species exist in a user-defined frame. For example, one could obtain species records for a rectangle defining the boundaries of a study site or a buffer created at a specified distance from a stream, a lake, or even a point where a related species is known to exist.

A modification of the latter technique changes the boundaries of the search area to conform more specifically to a given set of criteria. For example, a very useful search method might involve a determination of what taxa may be found in a particular habitat type. If these habitat type data are available, for example, through a GAP analysis or some other remote-sensing program, then these habitat polygons could be compared to point data indicating the species data for that polygon. This

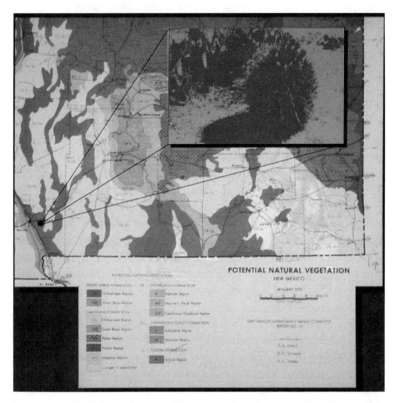

FIGURE 9.1 Illustration of the graphic capabilities of modern GIS software. This illustration shows a map of the natural vegetation of New Mexico, with one vegetation type highlighted and showing a blowup photograph of some common plants (a prickly pear cactus and a fishhook barrel cactus) shown in situ within that vegetation type. Base map modified from U.S. Department of Agriculture (1978). ARC/VIEW figure courtesy Environmental Systems Research Institute, Redlands, California.

type of search is called the "point-in-polygon" search and is a common, easily performed, and quite useful technique.

These point-in-polygon searches need not be limited to habitat types. Any existing data layer might be queried to show correspondences between its attributes and those of the species database. For example, you might want to survey the database to determine which sensitive, rare, or endangered species occur within a parcel of land that is being considered for a set-aside or easement. You might also like to search the database to determine the frequency of occurrence of a particular species on north- versus south-facing slopes. Or you could perform the same search on steep versus shallow slopes. The possible combinations are endless.

Beyond search-and-compare operations, however, GIS offers the user the ability to combine mapped coverages to create completely new coverages through its map-overlay operations. GIS creates the resultant maps through a process called overlay, but the process can be more complex than its simple name implies. In cell-by-cell-based GIS (also called raster-based), map coverages can be combined through any number of mathematical combinations of numbers assigned to represent map categories. These can be addition, subtraction, multiplication, division, squares, roots, diversity measures, minima, maxima, and so on. In vector-based GIS, the map-overlay operations normally involve set theoretic comparisons such as union, intersection, and complement. But it need not be limited to these. Maps produced from any one of these integrations of multiple coverages include complex polygons called integrated terrain units, or ITUs (Dangermond, Derrenbacher, and Harnden 1982). The ITU is a useful and well-thought-out integration of mutually associated phenomena. Defining areas on a map based on physical and biological needs of a taxon might be an excellent method for creating a sampling strategy.

Another powerful technique available in GIS that could prove highly useful in the preservation of phytodiversity is called "buffering." Buffering is a process of measuring a selected distance from a point, line, or polygon. This process can allow the scientist to determine a zone of protection in areas known to be sensitive. Or it might be used to further define a sampling scheme. The appropriate distance to buffer anything is a more difficult determination to make than is frequently recognized. Its correct determination may mean the success or failure of a protection strategy. A buffer zone may be defined based on the known distance that factory effluent contaminates the soils near its stream. Or a buffer distance may be applied around a protected area to preserve the characteristics of that area. Correct buffer size for either of these situations requires that some well-established empirical data indicate the correct size or that a well-founded theoretic measure of the correct distance be known.

Combining polygonal map coverages based on biological and physical taxon needs with polygonal coverages derived from buffer generations as well as with a coverage showing land ownership would create a useful set of ITUs. With this resultant map, for example, one could find areas that fit the prescribed needs of the taxon or taxa in question, that are well buffered from outside influences such as pesticides, and that are available for purchase. This set of polygons would then be prime candidates for purchase for species-protection strategies. A similar study could be conducted to show accessibility for field study and sampling. The sites could even be prioritized based on the degree to which certain criteria are met. If, for example, all the prescribed criteria meet at a certain polygon, this polygon is likely to be the highest-ranked polygon.

Providing a Geodetic Framework

As simple and straightforward as these GIS operations sound, two fundamental prerequisites are needed for their success. First, an appropriate grid system must be established for the collection, sensing, and locating of vegetative information; second, an appropriate map projection for the modeling effort must be chosen. A wide variety of grid systems are available, each designed for a specific purpose. The United States Public Land Survey (USPLS) system, for example, was originally designed to allow land to be allocated in large squared parcels. With its common township and range designation, it is easy to define locations, measure areas, and locate ownership boundaries, at least when the land is subdivided into these squarelike parcels. The state plane coordinate system is designed to allow for precise definition of land holdings, especially when they do not conform to the square parcel formats required by the USPLS system. A major drawback of this latter system is that it does not necessarily carry over from one state boundary to another, making matches between entities that cross these borders both difficult and frequently inaccurate. These latter systems may be useful for inputting GIS coverages specific to land-ownership data but are not very useful for field collecting of vegetative data. For that purpose it might be wiser to use the Universal Transverse Mercator (UTM) grid system. This system measures distances in meters north and east of preselected starting points called false northings and false eastings. Additionally, if some of your data have been collected by military establishments in the United States, it is probable that they will have been collected on the military grid, which is based largely on the UTM system.

Each region of the world and each nation has its own unique set of grid systems upon which that region or nation may have based previous sampling. It is important to define these grid systems, to understand their starting points and their particular advantages and disadvantages, so that new sampling might correspond with historical sampling. This is particularly important if the purpose of a study is to define the extent and direction of vegetative change for an area. For those studies that were based on aerial photography or on some other form of visual imagery, it is often necessary to obtain actual copies of the imagery. At the very least one should locate and define as many absolute coordinates as possible to coregister future coverages.

An adequate geodetic framework will also allow newly sampled specimens found in the field to be located accurately. Current Global Positioning System (GPS) technology allows for very exact locational identifiers to be attached to each sample location. The coordinates can include both X and Y coordinates, as well as topographic or Z coordinates. This will allow future researchers to identify not only the latitudinal and longitudinal positions of the species but also its absolute elevation. With these data it is probable that both positional and elevational conditions attendant to the species can be used for later predictive modeling of their whereabouts.

It would be very useful if the existing herbarium records contained precise locational identifiers. Unfortunately, these records are all too often written in descriptive terms, the meaning of which may change through time. It is possible, in some circumstances, to re-create some semblance of locational accuracy through appropriate combinations of linguistic analysis, historical archival analysis, and extrapolation to known biological limits of species to reproduce useful locational information within a GIS (McGranaghan and Webster 1988). Results from such analysis would need to be qualified so that any map coverages made that was based on them are well documented and the potential for inaccuracies stated clearly. When performed properly, such an historical vegetative database could prove helpful to biogeographical studies, especially those large-scale studies emphasizing regional species distributional changes, either in position or in area.

To examine large-scale changes in position or area it should be expressly stated here that the map projection used for data input from original cartographic data is very important to the success of any data comparison or analysis. All maps, whether in digital or analog form, are mathematical projections of a roughly spherical object, the earth. Even if we assume that the earth is perfectly spherical, no flat piece of paper can adequately represent such a surface without error and distortion. The amount and type of distortion are dependent on the type of map projection used. For example, equal area projections are designed to maintain the area of portions of the earth as displayed on a two-dimensional surface. To do so, however, the cartographer is required to sacrifice other properties, such as shape or distance (Robinson et al. 1984). Before beginning a project, it is best to determine which properties of the map must be maintained to model the vegetation patterns in question most accurately.

Although most modern GIS software offers a wide variety of map projections and the ability to alternate between different map-projection changes at will, it should be remembered that such manipulations come at a cost. Because each map projection is, by definition, an approximation of the already assumed spherical shape of the earth, mixing and matching map projections will cause overlayed coverages to lack compatibility in georeferencing, bordering regions to show gaps in coverages, and so on. Careful selection of map projections before data analysis will lessen the extent of these problems. Additionally, the smaller the region selected for a study, the less important the map projection becomes.

Conforming to Government Standards

Any project that is federally funded or that intends, as one of its goals, to provide data to federal agencies must conform to Federal Geographic Data Standards. For vegetation data, these standards are set by the Federal Geographic Data Committee (FGDC) Subcommittee on Vegetation Data (Federal Geographic Data Committee 1991). This subcommittee coordinates agency data interchange, compiles information on federal agency data, develops and evaluates data definitions and standards for national vegetation classification activities, determines appropriate vegetation categories, and promotes standards for vegetation data. Beyond merely conforming to these standards, however, those involved in the collection, preservation, and cataloging of phytodiversity information may influence the decisions made by the committee. Generally those who map vegetation data have little interaction with the curators of herbaria, and vice versa, and curators are seldom involved in the mapping process itself despite the large volumes of extremely useful information. Consequently, vegetation mapping standards are heavily biased toward nonherbarium-based information, thereby limiting the potential usefulness of herbarium samples for vegetation mapping. More interaction between these groups, it would seem, would benefit both.

A National Biodiversity GIS?

A number of obvious technical issues exist for the development of a national biodiversity GIS of which a phytodiversity GIS would be a part. It will be necessary to define the purpose or purposes of the national database. This will of course require the National Biological Service to determine who the users would be. Because of the bureaucratic nature of national agencies and the highly disparate nature of the potential users, it is very unlikely that all potential users will be identified before database development. The absence of this step will no doubt result in a lack of determination of appropriate map coverages, scales, sampling methods, and technologies necessary for many useful modeling tasks. A consequence of this oversight is likely to produce a large amount of data, all manipulated to conform to regional needs, with little cooperative effort from one region to the next.

Beyond the cooperative efforts necessary for the ultimate success of a national phytodiversity GIS, more pressing problems need to be addressed. Current vegetation sampling methods, for example, must be considered before a national phytodiversity GIS can become a reality. Detailed discussion of sampling procedures are discussed elsewhere in this volume. I would like to note, however, a persistent sampling problem that is of paramount importance to the spatial domain and to the ability of GIS to model global change. The problem is that of oversampling and undersampling.

Undersampling generally occurs because of the time and effort necessary for ground sampling, especially in remote locations. In addition, if the purpose of the sampling is to monitor change, the problem of undersampling becomes even more severe because of the necessity of repeat samples. Locations exemplifying rapid change are likely to be those receiving the least adequate repeat coverage. Spatially distinct species, that is, species that occur in clumps or in small patches, are also likely to be undersampled because of their small spatial extent. This will most likely happen when the species are known to be common. These patches may actually provide for a wide variety of species because they contain large amounts of both interior and edge species in a compact area. If there are known rare or endangered species in small patches, it is possible that these patches might be heavily sampled and therefore overrepresented in a herbarium collection.

Oversampling would not seem to present a problem unless one visits it in context with other sampling efforts. For example, species known to be commercially important are often sampled in great detail despite the low species abundance in such locations. In some cases large tracts of monocultures are sampled both spatially and temporally far beyond the norm. Alternatively, botanists are often aware of locations that exemplify very high species abundance. Such locations are extremely attractive because they allow for samples of many species to be collected within a reasonable amount of time and for minimal cost. In fact, such locations are often targeted by collectors from all over the world, thus highly exaggerating the importance of these species to the overall biota. What remains is a large number of moderately abundant species that are largely undersampled and therefore underrepresented in the world's biota.

Some possible solutions can avoid the trap of in situ sampling bias. First, the GAP maps currently being produced throughout the United States, as well as other similar remote-sensing projects designed to develop synoptic views of biodiversity, would seem to be most appropriately designed to produce a stratified sampling scheme. This or other sampling schemes could also be modified by the unit size of each given habitat type to improve the representative counts of species in large areas.

An increased temporal scale, such as that available with AVHRR data, could provide useful information about phenology, and about short-term habitat-condition changes necessary to improve sampling, by allowing the field teams to become aware of time-sensitive vegetation. This would be useful in sampling grasses, for example, because of the difficulties involved in identifying them when they are not in bloom. Moreover, it would be possible, with the use of ancillary data, to develop predictive models to determine sensitive areas or areas that should be sampled.

Beyond the sampling of specimens for input to GIS, existing herbarium collections present similar problems such as huge data volumes, taxonomic mistakes and changing classification systems, the structured hierarchical nature of classification systems, and, of course, missing, incorrect, or imprecise locational identifiers. Efforts to determine coordinate equivalents of linguistic locators have already been indicated

(McGranighan and Webster 1988); more research is needed, however, in identifying incorrect or missing locators.

The large volumes of data in herbaria do not pose an insurmountable problem. Today's gigabyte and terabyte storage units are already available at a reasonable cost, but a great deal of time will be needed to input these data into computer systems. The hierarchical nature of the classification system is not a serious problem either, given the powerful and flexible database management systems frequently tied to modern GIS. These systems are capable of organizing hierarchical data into tables with appropriate data keys allowing a multitude of search strategies. Yet these problems and their solutions are merely technical and pale by comparison with the larger design issues for implementing GIS of any size.

More often GIS has failed as a result of poor design than because of technical difficulties. A system of the size needed for a national biodiversity GIS is sure to fail if it is not thought out clearly. For this reason, Marble (1985) compiled a list of design considerations necessary to build GIS for preserving biodiversity on a national scale. His list suggests the following:

1. *Develop a data dictionary.* This is a detailed description of what is in the database, data sources, scales, projections, and any other pertinent information designed to allow subsequent users to become familiar with the data quickly and to know what all the terms mean.
2. *Determine individual user views.* This is a detailed description of what an individual would do with the system and its data in the day-to-day performance of his or her tasks.
3. *Establish modeling needs.* This generally advances the user's view by teaching the individual what the system is capable of doing and how it might enhance how the user currently performs his or her tasks.
4. *Perform data needs assessment.* Once the modeling capabilities are known, the necessary data, sources, scales, and so on, are easily identified.
5. *Perform user view integration.* This allows for a hierarchical classification of all user needs based on demands. High-demand operations will necessarily be put in a high-priority category, whereas operations needed by only a few or those merely suggested as "desirable" will probably not be included in the system implementation.
6. *Prototype GIS.* It is important to determine if the system will operate as designed. To do so a small subset of data is used to determine if GIS is capable of doing what is required.

7. *Test GIS performance.* In addition to determining GIS capabilities, the user must also evaluate whether the results will be consistent, whether they can be expected to perform in a timely manner, and whether the system is capable of handling the number of users required.

8. *Implement the final GIS.* If all the above measures have been implemented and are operating as planned, the last step is to implement the final GIS.

This is a simple list of requirements to develop a reasonable national biodiversity GIS. However, because all systems have a given life expectancy, as will this one, it is necessary to develop a flexible methodology to account for increased user awareness of GIS capabilities (and therefore increased user demands). The system will also become a victim of changing technology, varying sources of data, newer, more efficient software, perhaps even a GIS based on virtual reality. To ensure that the system can adapt to these changes, I suggest that a structured GIS design model, such as that proposed by Marble and Wilcox (1991), be considered before one attempts to create such a large system.

Literature Cited

Butler, K. S., A. K. Ludeke, and E. C. Palmer. 1990. Regional planning for endangered species protection using GIS. *Proceedings, GIS '90*, National Computer Graphics Association. Houston, Texas, August 26–29, pp. 104–110.

Dangermond, J., B. Derrenbacher, and E. Harnden. 1982. *Description of Techniques for Automation of Regional Natural Resource Inventories.* Environmental Systems Research Institute, Publication #220, 54 pp.

Davis, F. W., D. M. Stoms, J. E. Estes, J. Scepan, and J. M. Scott. 1990. An Information Systems Approach to the preservation of biological diversity. *Int. J. Geog. Inf. Systems* 4:55–78.

Federal Geographic Data Committee. 1991. *A National Geographic Information Resource: The Spatial Foundation of the Information-Based Society.* First Annual Report to the Director, Office of Management and Budget. Washington, D.C.: U.S. Government Printing Office.

Lillesand, T. M. and R. W. Kiefer. 1994. *Remote Sensing and Image Interpretation.* 3d ed. New York: Wiley.

Marble, D. F. 1985. *Geographic Information Systems as a Tool to Assist in the Preservation of Biological Diversity, Office of Technology, Congress of the United States.* Contract No. 533–4735.0.

Marble, D. F. and D. L. Wilcox. 1991. Measure twice-cut once: A structured approach to successful GIS design and implementation. *Proceedings,* 11th Annual Environmental Systems Research Institute User Conference.

McGranaghan, M. and L. Webster. 1988. Prototyping an herbarium collection mapping system. *Technical Papers*, 1988 ACSM-ASPRS Annual Convention: GIS, Vol. 5, pp. 232–238.

Robinson, A. H., R. D. Sale, J. L. Morrison, and P. C. Muehrcke. 1984. *Elements of Cartography*. 5th ed. New York: Wiley.

Scott, J. M, B. Csuti, K. Smith, J. E. Estes, and S. Caicco. 1990. Gap analysis of species richness and vegetation cover: An integrated conservation strategy. In K. A. Kohm, ed., *Balancing on the Brink of Extinction: A Retrospective on the Endangered Species Act*, pp. 282–297. Washington, D.C.: Island Press.

Scott, J. M., F. Davis, B. Csuti, R. Noss, B. Butterfield, S. Caicco, C. Groves, J. Ulliman, H. Anderson, F. D'Erchia, and R. G. Wright. 1993. Gap analysis: A geographic approach to protection of biological diversity. *Wildlife Monographs*, no. 123, pp. 4–41.

Stoms, D. M. and J. E. Estes. 1993. A remote sensing research agenda for mapping and monitoring biodiversity. *Intl. J. Remote Sensing* 14:1839–1860.

U.S. Congress, Office of Technology Assessment. 1987. *Technologies to Maintain Biological Diversity*, OTA-F-330. Washington, D.C.: U.S. Government Printing Office.

U.S. Department of Agriculture. 1978. *Potential Natural Vegetation: New Mexico*. Soil Conservation Service, New Mexico Interagency Range Committee, Report No. 11.

PART 4

Preservation of Materials

Since the sixteenth century, botanists have been making herbarium collections that have resulted in 270 million specimens housed in the world's institutions (Baum, chapter 4). The value of these materials has been proven many times over, and they serve as a basis for most of our botanical systematic publications. But these are not the only materials that have been made from plant parts.

As Sherwin Carlquist outlines in chapter 10, wood samples have been collected in the field and stored in xylaria, usually dried but sometimes pickled. Such samples are useful for detailed cytological, developmental, or embryological studies, but they require greater effort to acquire and maintain. New types of plastic bottles, lids, and seals, however, have lessened this curatorial burden. Pollen is a similar type of sample and represents similar types of problems, but the herbarium sheet itself with fertile material serves adequately for many palynological needs. The critical questions are the degree to which we want to continue these types of additional samplings and their proper curation, and if so, who will take responsibility.

Eric Roos and colleagues in chapter 11 further address the topic of preserving plant parts, in this case seeds, pollen, and buds. They discuss low-temperature storage, a methodology that has reached high levels of sophistication. The question remains as to which institutions and what herbaria, if any, should be pursuing these types of preservation methods routinely. Another concern is taxic representation, recognizing that not all species can be preserved in such a relatively costly manner. These are priorities that must be established.

Chapters 12 and 13 deal with DNA and its utility in systematic work. Dennis Loockerman and Robert Jansen discuss issues relating to the use of DNA from herbarium material—of great importance considering there are 270 million packages of DNA stored in institutions worldwide. As their survey conveys, most herbarium curators are sympathetic to having DNA removed for particular research needs, but only within certain limits. Because of degradation owing to time and particular preservation techniques (chemical immersion, heat, etc.), some of the DNA has been already degraded; nonetheless, much still can be gained from examining what does remain. Robert Adams in chapter 13 stresses the importance of cold storage of DNA taken from fresh material. This is high-quality DNA, which is most valuable for comparing entire genomes as well as for serving as useful gene sequences for crop improvement and other economic and applied purposes. The extent to which plant collectors and herbaria should participate actively in these efforts needs to be studied carefully. DNA could routinely be collected during many expeditions if a community priority were placed on this initiative.

10

Materials for Anatomical and Ultrastructural Studies: Pollen, Wood, and Pickled Materials

SHERWIN CARLQUIST

Despite the drive to collect more plant specimens as natural areas dwindle, the rate at which wood specimens have been added to wood collections (xylaria) has decreased. Especially poorly represented are wood samples of shrubs, lianas, and woody herbs. Most xylarium samples lack bark (rich in systematic features), and wood of roots is almost never collected. Nondestructive methods of wood collection appropriate to current and future restraints on collecting are presented and evaluated. Pickled specimens are usually collected for a single purpose by or for a particular worker. If these collections are of use to more than a single worker, they should be shared. Long-term storage of pickled material is urged for type material, endangered species, and so on, but the value of a general pickled collection should be assessed on a case-by-case basis. Currently available plastic containers and silicone sealants permit maintenance-free storage of pickled material for indefinite periods, so the need for maintenance should not be advanced as a reason for abandonment of pickled specimens.

Herbarium specimens will continue to be a primary source of pollen specimens, although palynologists should work from liquid-preserved material where possible. The use of herbarium specimens for plant anatomy and ultrastructure is endorsed, but with appropriate safeguards to the integrity of these collections.All plant material collected in some form other than herbarium specimens should be accompanied by voucher specimens.

Although plant anatomy as a discipline is now more than three hundred years old, only a small proportion of vascular plants have been studied in any detail with respect to anatomy and ultrastructure. There are many plant-structure character states of interest with respect to ecology, systematics, and other fields, so that study of plant structure in the future will be highly desirable. Although the literature of systematic anatomy/ultrastructure appears extensive, the anatomy of perhaps only 5% of flowering plants is currently known in any detail. Field botanists tend to collect only herbarium specimens. Plant anatomists tend to collect for particular projects, at the conclusion of which their materials are often discarded. Only a relatively small number of botanists have collected materials for anatomical studies at large, e.g., collecting wood samples along with the preparation of herbarium specimens. Citation of these individuals here would be invidious, but the most significant names can, for example, be found in descriptions of the holdings of various xylaria (Stern 1988). One extensive manual on collecting plants and animals (Knudsen 1972) offers no instructions for collecting anything other than herbarium specimens of vascular plants, even though it was written in a period when anatomical studies were more commonly pursued than they are today.

Wood Collection and Preservation

Most systematists are unaware of the existence of large collections of xylaria, although these have been listed in a useful publication, *Index Xylariorum* (Stern 1988). The largest of these collections is that of the U.S. Forest Products Laboratory in Madison, Wisconsin, although many collections exist (Stern lists 134 of them but many of those are insignificant). Xylaria have multiple applications (Stern 1976) and are useful in much wider enterprises than simply systematic botany, for example, ecology, paleobotany, and wood technology. Yet these collections are not being increased as rapidly now as they were in the past when Stern (1973) discussed their remarkable potential.

Xylaria were mostly begun as adjuncts to forestry schools, where they were used for the study of wood properties. Not surprisingly, specimens were prepared as small boards (6"x 3"x1/2"), which could be conveniently stored in drawers. These boards have the disadvantage of lacking bark, of being from unspecified parts of a plant, and of being from plants of unknown age. Today, xylaria more frequently add specimens that are cylindrical portions of a branch or sometimes a basal stem. These portions occasionally do have bark. Bark is rich in systematic characters. For example, bark anatomy was more decisive than wood anatomy or many other features in demonstrating the affinity of the new family Ticodendraceae to coryloid Betulaceae (Carlquist 1991a), and bark provides excellent evidence for the monophyly of Gnetales (Carlquist, unpublished). Bark is rarely present on herbarium specimens, so to the degree that it is not represented in xylaria, we have inadequate collections of this important feature.

Xylaria have other significant biases, of which users should be cognizant. Hopefully these biases will eventually be diminished. Xylarium specimens are almost all of stems, although roots have wood significantly different from that of stems and are worth attention, as one pioneering study (Cutler et al. 1987) has demonstrated. Xylarium contents are heavily biased toward tree species, not surprisingly since most xylaria were begun in forestry schools or institutes. Woods of lianas, shrubs, woody herbs, and woody succulents are markedly underrepresented in xylaria. Workers who have studied wood in these categories have had to rely largely on wood they collected or wood collected especially for them. A surprisingly large number of annual flowering plants have wood sufficient for study, but xylaria virtually without exception lack wood of annuals; herbarium specimens also have suboptimal wood on specimens of annuals, because collectors usually avoid the larger individuals (which would provide more wood).

Today, relatively few individuals are engaged in collecting wood samples, and most of these individuals are interested in forest exploitation rather than preservation of systematic diversity. The largest wood collection (U.S. Forest Products Laboratory at Madison, Wisconsin) had (as of 1988) 35,000 wood samples. This seems large until one realizes that many common species are represented by multiple specimens, especially species with commercially usable woods. The number cited also seems small in comparison with herbarium specimens, since close to a hundred of the world's herbaria have more than a million specimens.

Xylaria in general are willing to fill requests for small sectioning blocks from qualified workers in wood anatomy. Workers request information concerning holdings of a particular genus or family, and curators of larger xylaria typically send photocopies of index cards from which the worker may request sectioning blocks. Computerized printouts will doubtless be supplied in years to come. From their specimens, curators of xylaria are usually willing to remove sectioning blocks, each usually about 1 cm in each dimension. A wood-section slide is generally expected in return. Curators of xylaria are also willing to lend such slides. Requests for sectioning blocks have increased over the years, and xylaria may soon be expected to be more restrictive in the future.

Those who obtain sectioning blocks or slides from xylaria should expect modest documentation. Herbarium voucher documentation of woods is good in more recent specimens, but it is poor in specimens collected before 1950 (Stern and Chambers 1960). As with herbarium specimens, there may be weaknesses in accuracy of determination and nomenclature (Baas 1980). My experience has been that perhaps about one in a hundred wood samples is incorrectly identified as to family or genus; misdeterminations at the species level are no doubt more frequent but are difficult to uncover because differences among species of a genus in wood anatomy are often not dramatic. This may not seem a serious matter until one realizes that published information may be based on a misidentified specimen. For example, Rao et al. (1992) showed that the wood description of *Thottea* (and *Apama*) (Aristolochiaceae), as well as the familial description for Aristolochiaceae in Metcalfe and Chalk (1950), was based on an incorrectly identified xylarium sample, which appeared to be lauraceous. Woods of lianas in xylaria in my experience are particularly likely to be misidentified: connecting the wood taken from near ground level with flowering parts in the canopy is difficult. Wood misidentifications are difficult to remedy: one cannot readily determine the species or even genus with a slide of wood sections from the sample, whereas determining a genus and species of a herbarium specimen is a comparatively easy task. Thus the number of misidentifications in xylaria is likely to persist, whereas in herbaria, as groups are monographed, identifications improve. The best way to test a suspected misidentification of a xylarium sample is to compare sections of it with sections of a twig from a correctly determined herbarium specimen of the species in question.

Long-term preservation of wood samples is not difficult in temperate areas. Pith and softer wood samples may be attacked by wood-boring beetles or even dermestids. Although wood samples are much more resistant to insect damage than are herbarium specimens, occasional fumigation is necessary. In tropical areas, a wide range of boring insects, most notably termites, pose a greater threat. The reality of this threat has been demonstrated by the fact that one wood collection in a humid subtropical city was destroyed to a large extent because that collection was stored in a building with unscreened louvered windows. Damage to wood samples by fungi is unlikely except in the instances of soft woods in very humid tropical climates.

Collection of wood samples for xylaria at a time when species are vanishing seems a prime desideratum, but this is being largely neglected in industrialized countries in comparison with greater activity in earlier years. Interest in forest exploitation in some nonindustrialized countries has led to collection of wood samples, but many of these newer collections are poor in woody species not considered of commercial value. Most xylaria contain tree species predominantly; some xylaria contain arboreal woods exclusively. Species of shrubs, lianas, and woody herbs are notably underrepresented despite the systematic and evolutionary interest of such woods: the fact that houses and furniture cannot be constructed from wood of shrubs accounts for bias against such wood in many wood collections. Therefore, workers interested in wood of nonarboreal species will have to collect their own materials to a large extent.

If impetus for collection of wood samples has lessened in recent years, the difficulties have not. The traditional means of harvesting a wood sample, i.e., cutting down an entire tree and taking a portion of the basal stem, is now considered wasteful and unacceptable. Selection of smaller portions without destroying an entire tree is usually feasible. A survey of nondestructive means of wood sampling is offered by Swart (1980) and Echols and Mergen (1955). Some methods that seem plausible are not; for example, corings used for age determination or dating are too small in diameter and are wrong in shape for sectioning. Twigs or slender branches are often selected by well-meaning individuals supplying wood samples to a colleague, but samples of small diameter are relatively juvenile, and therefore data from them is dubiously comparable to data from wood of a mature stem (or root). One method that has served me well was introduced to me by Japanese botanists with whom I traveled: one looks for a branch that recently has died but has not yet been attacked by fungi. A lower branch with few leaves, shaded by newer upper branches, is also a good selection, a kind of judicious pruning. Attention is also called to the possibility of collecting wood in areas of road construction or other development. Recently, the efforts of Jack Fisher and volunteers at the Fairchild Tropical Garden resulted in preservation of wood samples from trees, shrubs, and lianas severely damaged in 1992 by Hurricane Andrew. Both stem and root wood of most species was preserved. A listing of these specimens has been issued by Fairchild Tropical Garden, and portions suitable for anatomical study are available upon request to wood anatomists.

Preparation of wood samples from a living stem or root provides difficulties greater than those encountered in preparing herbarium specimens. The areas richest in woody species are humid tropical or temperate forests where drying of woods without excessive molding is difficult. A wood sample that has been invaded extensively by fungi is of limited value for anatomical studies. I ordinarily avoid such samples unless they are of unique systematic value (e.g., the only sample available for a particular genus). The remedy used for herbarium specimens—application of heat to achieve drying before fungal infestation—is of limited value for wood samples. When dried under artificial heat (except where done by particular methods), wood samples tend to develop numerous small cracks ("checking") that interfere with sectioning. Air drying in a place with constant passage of air (e.g., a rooftop protected by a simple shelter to exclude rain) may be the simplest expedient. Removal of bark facilitates drying but results in the loss of a systematically valuable structure. Removal of bark and even pith before drying is the only method other than pickling that can serve for soft woods vulnerable to attack by fungal and bacterial rot. For long-term storage, drying is the most satisfactory method of sample preservation. Little anatomical information is lost by drying woods. If one needs information on the presence or absence of nuclei in wood fibers, pickling in 50% aqueous ethanol is desirable.

Drying of wood samples may be more easily accomplished at the collector's home institution rather than in the field. If so, wood samples can be shipped in some chemical that deters fungal or bacterial action. The chemical can be removed by washing when woods arrive at the home institution, and drying can proceed. Ethanol is perhaps the best chemical in which wood samples can be preserved during the shipping process, but ethanol is not readily available in many countries. Methanol is an entirely satisfactory substitute for this purpose.

Paraformaldehyde powder is light and can be readily carried by a collector; formalin (formaldehyde in water solution) is often available in many countries. Formaldehyde is, however, a noxious and unpleasant chemical and is therefore less desirable than ethanol or methanol to prevent rotting. Samples can be shipped in wide-mouth plastic bottles, plastic tubs (e.g., Tupperware®), or heavy-gauge plastic bags capable of leakproof closure. A plant anatomist receiving such samples can remove pickled wood and bark for anatomical study, then dry the remaining portions for inclusion in a xylarium. The simplest fluids suffice for most purposes (e.g., dilute ethanol is just as good as formalin-acetic-alcohol). If a worker is doing transmission electron microscopy (TEM) studies on living wood cells, more specialized fixation techniques are desirable, but those techniques are not applicable to larger wood samples and should not be used for general purposes.

Slides of wood sections (traditionally, a transverse, a tangential, and a radial section of a sample mounted on each slide) are often kept as adjuncts to wood samples in xylaria. Such slides can be loaned, and by loaning a slide instead of supplying a portion for sectioning, a xylarium can conserve its wood samples. Because one can easily produce several sections instead of just one, and therefore make replicate wood-section slides, wood anatomists should be encouraged to prepare sets of slides. These can be exchanged with other workers or xylaria. I have found replicate slides useful because one slide of such a set can be soaked in xylene (or another solvent), the resin removed, and the section studied with scanning electron microscopy (SEM). Thus my wood-section slides were a library that supplied sections for SEM studies on innumerable occasions.

Xylaria are bulky and require special curation. If a worker who has accumulated a xylarium leaves an institution and that institution does not continue interest in wood studies, consideration should be given to transferring the xylarium (and its wood-section slide collection) to an institution where wood studies are actively occurring or where good curation is likely to make the materials readily available to workers at large.

Pickled Materials

Almost every institution that has supported a botany program for several decades or more has, in one or more places, a collection of pickled material. Some of these may have been used in class demonstrations, some for research. Should one keep such collections? If so, what is the best method for long-term storage with a minimum of maintenance? Should one keep research pickles once the research has been accomplished? Should one keep a pickled collection as a herbarium adjunct much as cone or fruit collections are often kept?

The answers to these questions depend both on the ease of maintenance of pickles and on the usefulness of the material. Researchers have often obtained materials with great cost and effort. If unusual or rare materials were obtained for, say, a study on embryology, an effort should probably be made to see if a student of floral anatomy or floral development would be interested in taking them. If the pickled materials have been adequately studied for a particular purpose, do not represent rare or endangered species, and are not wanted for instruction or by other researchers, they should likely be discarded. In fact, I truly cannot support the use of pickled materials for teaching—they are, in my experience, unappealing visually and are a poor way of presenting examples of particular plants or structures to students. The problems of removing a plant portion from a jar, displaying it in a tray while keeping it moist, and then returning it to the jar make the use of photographs or herbarium specimens more feasible.

Pickled material of type specimens is of greater than ordinary value. Those who deal in taxonomy of groups of succulents may find pickled material useful for optimal preservation of shape of structures. Pickled material of rare or endangered species seems of great potential value. However, the problem inherent in a pickled collection is that of making it available to potential users. Although the existence of xylarium samples and herbarium specimens can be readily ascertained by interested workers, pickled collections are not ordinarily cataloged, and no listing exists of pickled collections and their contents. Thus I feel the owner of a pickled collection has the responsibility to find potential users for the material, and if no such users can be found while most containers still contain liquid, the material should be discarded. If maintenance of a pickled collection is not feasible but the material is of interest, one can convert pickled plant portions into herbarium specimens. One could present plausible reasons for maintenance of a few centralized collections of pickled material, but major institutions seem at present unlikely to take on such a responsibility.

Today, the question of maintenance of pickled collections has become less troublesome because of the existence of better containers. Glass jars with glass lids and rubber sealers, invented for food preservation, require periodic replacement of the rubber sealers. Wide-mouth glass jars with bakelite screw-cap lids require periodic replacement of the lids, because components of pickling solutions cause warping and even cracking of the bakelite with consequent evaporation of liquid. Paper or cardboard sealers in such lids tend to shrink and permit evaporation, even if the lids are periodically tightened. Although long-term experience with plastics such as polystyrene, polyethylene, and vinyl is not available, bottles and lids made of some of these plastics (under brand names such as Nalgene® and Tupperware®) or glass bottles with lids made of some of these plastics seem likely to have long-term persistence of the plastic without warping or cracking. Once tightly closed, the lids of these bottles appear to remain seated for long periods so that looseness of lids is not a problem. However, for long-term storage, I recommend applying a silicone sealant. (I have used a type that has acetic acid as a solvent.) Plastic tubs of pickled material that I sealed in this way in South Africa in 1973 survived being shipped by sea mail and have shown little or no evaporation for the past twenty years. The disadvantage of most plastic containers is their opacity, requiring removal of the material to ascertain the nature of the contents. There are some clear plastic containers available, but thus far these appear not to be used for pickled material because they are relatively brittle and may not seal as well as the containers made from opaque plastics. The potential longevity and freedom from leakage and evaporation of currently available plastic bottles and tubs makes them much more satisfactory than containers used in past decades.

Plant morphologists typically pickle plant portions in formalin-acetic-alcohol. This is indeed a proved fixative for light microscope work, but if excellence of fine histological or cytological details is not of great significance, preservation in 50% aqueous ethanol is entirely adequate. Such ethanol is readily available in U.S. laboratories, but may not be obtainable in foreign countries except in the form of alcoholic beverages. The first material collected for anatomical studies of *Degeneria* was pickled in gin (A. C. Smith, personal communication). When traveling in foreign countries, I have found gin and vodka entirely satisfactory as fixatives. Pickled material stored in an alcoholic solution need not have additional quantities of that precise fixative added to compensate for evaporation—the addition of 50% aqueous alcohol should be entirely satisfactory.

Embryological and developmental data cannot be obtained from dried specimens, but virtually every other kind of anatomical information can be obtained from herbarium specimens. If one wishes comparative data on leaf anatomy or other vegetative structures, herbarium specimens—even old ones—are entirely adequate, with a few exceptions. Dried material of a few families—notably Solanaceae—does not yield to expansion in such a solution as 2.5% aqueous NaOH. As noted above, mature wood patterns of wood anatomy are not available from twigs of herbarium specimens of woody plants. Keeping these exceptions in mind, herbarium specimens can serve for anatomical studies that include not merely sections of vegetative, floral, fruit, and seed material but also studies via SEM of wax secretions on leaves, epidermal relief, and trichomes, as well as clearings to reveal leaf venation. Thin epidermal cell walls and trichome walls can make herbarium specimens less desirable for SEM studies than critical-point-dried portions of liquid-preserved materials, but where walls do not collapse with drying, as in leaves of Bruniaceae (Carlquist 1991b), dried materials yield excellent results.

The above remarks may seem a deviation from the topic of collections of pickled material. However, my intention is to show that dried materials may be adequate for many purposes, and this potentially limits the need for liquid-preserved specimens. The desirability of liquid-preserved specimens should therefore be viewed in light of the information that can only be obtained, or obtained best, from pickled specimens. Pickled material is often thought to be the entire basis for studies in plant anatomy, but increasingly such studies depend on pickled material supplemented by herbarium material.

Pollen

For TEM studies of pollen, workers ideally would like glutaraldehyde fixation. Such techniques are applied where ultrastructural and cytological details are to be studied but are not required for comparative studies of pollen exines. Exine, composed of sporopollenin, a carotene polymer, is extraordinarily resistant, accounting for the presence of well-preserved pollen grains in fossil deposits. Dried pollen grains, provided they are suitably treated, can yield reliable exine information, although liquid-preserved grains are more likely to yield more accurate information on pollen grain shape. Dried grains tend to experience elongation in the polar axis with drying, and indentation of apertures or even portions of the wall other than apertures may result from drying. Grains with very thick walls and more rigid pore coverings, as in Asteraceae, show very little alteration by drying. In a population of grains that have been dried, a certain proportion may look relatively natural in shape; also, there are methods of treatment that are likely to expand grains well, as described in the many papers featuring numerous SEM photographs of grains derived from dried specimens.

The nature of layers of exine is, in the opinion of some workers, actually seen best not in sections viewed by TEM but in SEM photographs of exines that have been fractured (Skvarla, Rowley, and Chissoe 1988). Methods for preparing grains for SEM study of exposed exine stratigraphy are detailed by Skvarla et al. (1988).

If, as the above indicates, dried pollen grains can serve adequately for most pollen studies, how should one store a collection of dried pollen samples? The anthers of a flower or a male conifer cone are ideal devices for retention of pollen grains, with the considerable proviso that once opened, anthers may retain few grains. Therefore, what one wishes are anthers of flowers in bud, just before anthesis. Pollen from young buds substantially before anthesis may have immature exine structure.

One could imagine a collection of dried flower buds in envelopes as a source of pollen material. However, herbarium specimens come close to matching such a hypothetical collection; in addition, herbarium specimens bear vegetative parts that can aid in checking the determination of the specimen. Herbaria have therefore been used as a source of pollen grains for many studies, supplemented by liquid-preserved grains where available or desirable.

Herbarium Specimens: Should Plant Anatomists Exploit Them?

If pollen workers removed pollen-bearing flowers, or even a few stamens, from a herbarium specimen with few flowers or flowers with few or large stamens, potential harm could be done to the specimen. Certainly removal of a small quantity of material is not likely to harm a specimen appreciably. Pollen workers would be well advised to obtain liquid-preserved material where possible, because more information is available from the study of pickled grains. Similar considerations apply to anatomical studies on sectional leaf anatomy, stem anatomy, floral anatomy, fruit and seed wall structure, trichomes, and leaf venation.

The potential demand for herbarium specimens for such studies could cause curators to restrict access to such specimens.

Restricting access to herbarium specimens does not seem to me a tactic of value to science any more than refusal by a librarian to allow use of books. An alternative strategy would be for field botanists to recognize the many uses to which herbarium specimens can legitimately be put, and to collect more numerous specimens with that in mind—indeed, many field botanists do collect in large sets. I see no harm, where rare species are concerned, in having two or three replicate sheets of a particular collection in a given herbarium. I assume in such a case that replicates will be sent to other herbaria as well as retained by the collector's institution.

I believe that harbarium curators would be well advised to respond to requests for portions in a way that permits legitimate study while minimizing damage to a specimen. Some friendly counsel by a curator, together with written guidelines, should suffice. In addition, annotation labels that indicate removal of parts by a particular worker for a given study can be affixed to a herbarium specimen. Such annotation labels are a good idea for several reasons. The labels would alert botanists to the existence of anatomical or ultrastructural information based on a particular specimen. Annotation labels also would tend to ensure sensible removal of portions, since a worker is certifying with the label that some parts have been removed. Herbaria should be encouraged to take the lead by providing annotation labels on which a worker can fill in the name and nature of the study.

Most curators would agree with the proposition that a herbarium specimen never studied by anyone is of minimal value to science (until it *is* studied). How much usage of herbarium specimens is valid, and what kind of usages should be permitted or even encouraged, are sensitive questions, however. I would like to emphasize that a specimen increases in value when information has been derived from it—and the more information derived, the more valuable the specimen. A specimen that has not been used in any kind of study is literally without value until it has served as a source for published information, at which time it acquires value commensurate with the published information. This assumes that the specimen has not been seriously degraded to the point of loss of diagnostic features by removal of portions. An example of how valuable a herbarium specimen can be when studied by various workers is shown by the great amount of anatomical information derived from the single collection of *Takhtajania* (Winteraceae), a

Madagascar genus that has not yet been rediscovered (see Leroy 1993; some interpretations of floral material are open to question, and only more material can clarify these).

Herbarium specimens have been traditionally regarded as the basis for floras and revisions, and emphasis on this view could lead curators to view plant anatomists as a potential threat to a collection rather than individuals who impart greater significance to specimens. The tendency to equate the use of herbarium specimens with taxonomic revisions, based on gross morphology, is demonstrated by the custom of loaning specimens to monographers with the requirement that identifications be verified and annotation labels attesting to determinations be affixed. Loan of specimens for an anatomical study that does not also involve determination of specimens is currently not customary, although I think it should be. Loans specifically designed for anatomical or ultrastructural studies, with appropriate curatorial guidelines, are likely to minimize damage of specimens, because a returned loan is subject to examination by the loaning curator. There is no point in pretending that removal of portions for anatomical studies from herbarium specimens is not going to occur. It will indeed occur, so the question is how to obtain responsible usage for these purposes.

In fact, many plant anatomists contribute to herbaria by donating specimens as vouchers for their studies, studies that involve the collection of material (e.g., wood samples) in addition to herbarium specimens. One must continue to stress that the preparation of voucher specimens for anatomical work is essential (Stern and Chambers 1960; Baas 1980) and that dictum should apply to any new kind of comparative study (e.g., DNA studies) where a correct identification is of value. To be sure, determinations on voucher specimens are not often checked by subsequent workers, although the existence of the specimens permits such checking at any time. More important, a worker in, say, comparative plant anatomy concerned about the preparation of accompanying herbarium specimens would seem more likely to determine species correctly than an anatomist who does not bother to prepare such vouchers.

Literature Cited

Baas, P. 1980. Reliability and citation of wood specimens. *IAWA Bull.* 2 (1): 72.
Carlquist, S. 1991a. Wood and bark anatomy of *Ticodendron*: Comments on relationships. *Ann. Missouri Bot. Gard.* 78:96–104.
———. 1991b. Leaf anatomy of Bruniaceae: Ecological, systematic, and physiological aspects. *Bot. J. Linnean Soc.* 107:1–34.

Cutler, D. F., P. J. Rudall, P. E. Gasson, and R. M. O. Gale. 1987. *Root Identification Manual of Trees and Shrubs*. London: Chapman and Hall.

Echols, R. M. and F. Mergen. 1955. How to extract large wood samples from living trees. *J. Forestry* 53:136.

Knudsen, J. W. 1972. *Collecting and Preserving Plants and Animals*. New York: Harper and Row.

Leroy, J.-F. 1993. *Origine et Évolution des Plantes à Fleurs*. Paris: Masson.

Metcalfe, C. R. and L. Chalk. 1950. *Anatomy of the Dicotyledons*. Oxford: Clarendon.

Rao, R. V., R. Dayal, B. L. Sharma, and L. Chauhan. 1992. Reinvestigation of the wood structure of *Thottea siliquosa* (Aristolochiaceae). *IAWA Bull.* 2 (13): 17–20.

Skvarla, J. J., J. R. Rowley, and W. F. Chissoe. 1988. Adaptability of scanning electron microscopy to studies of pollen morphology. *Aliso* 12:119–175.

Stern, W. L. 1973. The wood collection—what should its future be? *Arnoldia* 33:67–80.

——. 1976. Multiple uses of institutional wood collections. *Curator* 19:265–270.

——. 1988. Index Xylariorum 3. *IAWA Bull.* 2 (9): 203–252.

Stern, W. L. and K. L. Chambers. 1960. The citation of wood specimens and herbarium vouchers in anatomical research. *Taxon* 9:7–13.

Swart, J. P. J. 1980. Non-destructive wood sampling methods from living trees: A literature survey. *IAWA Bull.* 2 (1): 42.

11

Preservation Techniques for Extending the Longevity of Plant Tissues

ERIC E. ROOS, LEIGH E. TOWILL,
CHRISTINA T. WALTERS, SHEILA A. BLACKMAN,
AND PHILIP C. STANWOOD

Many countries have either developed or participate cooperatively in genetic resources conservation programs. In the United States, the USDA-ARS has developed a National Plant Germplasm System to acquire, preserve, evaluate, and distribute germplasm of some 8,250 agriculturally important species. These species are represented by more than 420,000 unique accessions, mostly stored in the form of seeds, but including live plants growing under greenhouse, screenhouse, botanic garden-arboreta, or field-orchard conditions. Long-term preservation of these materials, for use by future generations, is the primary mission of the National Seed Storage Laboratory in Fort Collins, Colorado. The technologies to accomplish this mission generally involve desiccation and storage at low temperature, including cryogenic storage in -196°C or over -160°C liquid nitrogen (LN). Survival of some seeds for hundreds of years has been reported for several species, and in some instances claims have been made for seed survival after thousands of years. These reports and claims undoubtedly represent the longest known viability periods for living tissues. For nonliving tissue we now know that at least the DNA can remain partially intact for millions of years. Storage protocols for seeds have recently been reexamined in light of new information on the mechanisms of seed deterioration and the role of water in stabilizing tissues during storage. Attempts to predict seed longevity have relied on extrapolating data from seeds stored under less than ideal storage conditions and thus have not satisfied our precise needs. Rescue of deteriorated seeds via embryo excision and tissue culture can provide a mechanism to maintain genetic diversity of some materials.

Storage of tissues other than seeds, such as pollen, dormant buds, and shoot tips, has now been demonstrated for a wide range of plant species. In the case of pollen, storage periods of one or more decades have been reported. The technology is similar to that for seeds in that longevity depends on lowering the moisture content and the storage temperature. Cryopreservation of dormant buds, harvested in midwinter, requires partial desiccation in the laboratory before cooling to the temperature of LN. Apple buds have been successfully stored for more than four years. Other species that can cold-acclimate may also be candidates for this storage procedure. Tissues with high water content, such as shoot tips, require specialized protocols that allow for the placing of germplasm materials in LN for storage and eventual retrieval. Different species and sometimes different cultivars of the same species may require unique procedures. These include the use of various cryoprotectants in different concentrations.

Germplasm Preservation Programs

Thousands of plant species have developed through natural evolutionary processes. Only a small percentage of these has been selected and utilized as agronomic, horticultural, ornamental, or forestry crops. Most plant genetic resources exist in natural ecosystems according to the principle "survival of the fittest" with no inventory and no managed preservation. As the world's population continues to expand, the areas required for intensive agriculture and forestry will increase at the expense of habitats for other plant species. Because in situ preservation will not be entirely adequate in the future, ex situ preservation must be expanded. In this chapter we examine the various strategies available for long-term ex situ preservation of seeds and other forms of plant germplasm.

Preservation of germplasm will only be accomplished if the various national and institutional infrastructures are sufficient to support the great challenge before us of conserving biodiversity. Historically, support for genetic resources programs has been minimal, at best. Worldwide interest in germplasm collection and preservation dates back to about 1961 when the Food and Agriculture Organization (FAO) of the United Nations held a meeting in Rome to discuss the general subject of plant exploration and conservation. A much broader meeting was held in 1967, again under the auspices of FAO with the cooperation and input of the International Biological Program (IBP) (Bennett 1968). The 1967 conference attempted to "develop approaches and methods, clarify contentious issues, point to areas in need of further research, and, above all, stimulate widespread interest and involvement" (Frankel 1970).

More than twenty-five years later, we are still grappling with these same issues, but our urgency is now much greater. Significant progress has been achieved in the last two decades on at least some of the more technical issues of preservation, and many countries have established strong germplasm programs and gene banks for conservation (see Baum, chapter 4). Technologies for conservation of plant tissues have been developed and modified in such a way that much of our agricultural germplasm can be preserved for extended periods, in the form of seeds, if properly grown, harvested, and prepared for long-term storage. Even some recalcitrant seeds (those that cannot survive desiccation) may now be preserved using a newly developed protocol. Nonseed germplasm such as pollen, somatic embryos, dormant buds, and shoot-tip meristems can also be preserved, usually in liquid nitrogen (LN); however, special protocols need to be developed and followed. Before examining these technologies in detail, a brief description of the international efforts on germplasm preservation and, more specifically, those of the United States is appropriate.

International Programs

Numerous governmental and nongovernmental agencies, companies, groups, institutes, and organizations are involved in attempting to slow or reverse the rate of loss of biodiversity around the world. It is not possible, nor is it the intent of this chapter to review all these efforts. However, the important work and contributions of two organizations are briefly described in order to underline the immense importance of this issue to the security of the world's food production.

CGIAR   The Consultative Group on International Agricultural Research, or CGIAR, is an informal association of forty public and private-sector donors that supports a network of sixteen international agricultural research centers (chapter 4, fig. 4.9). Established in 1971, CGIAR works in partnership with national research organizations to accomplish its mission of sustainable agriculture, forestry, and fisheries production in developing countries. Most of these centers maintain germplasm for the crops under their responsibility. The centers holding plant genetic resources account for more than 500,000 individual accessions of farmer varieties, wild species, and developed varieties or populations.

One of the sixteen research centers of particular interest to germplasm curators is IPGRI, the International Plant Genetic Resources Institute (formerly IBPGR, International Board for Plant Genetic Resources). IPGRI's current mandate is to advance the conservation and use of plant genetic resources for the benefit of present and future generations (IBPGR 1993). No other institution has done so much to promote the establishment of gene banks around the world. By 1989, fifteen years after its inception, IBPGR had fostered the establishment, development, and coordination of more than a hundred gene banks for the long-term preservation of seed germplasm. Plant explorations, funded in whole or in part by IBPGR, have resulted in the collection of nearly 200,000 accessions of plant germplasm. In addition, IBPGR has funded research on plant germplasm preservation of seeds and other forms of plant propagules. This research has resulted in greatly increased seed longevity and preservation protocols for many plant tissues. This work, along with the research conducted at the USDA National Seed Storage Laboratory (NSSL) in Fort Collins, Colorado, and the work of many others, has led to the development of international standards for gene banks to improve the long-term conservation of seed germplasm (FAO/IBPGR 1992).

The network of gene banks established by IBPGR/IPGRI has recently been reassigned to the jurisdiction of the Food and Agricultural Organization of the United Nations (see below). Under its new name, IPGRI will focus its attention on four major objectives: (1) to assist countries, particularly in the developing world, to assess and meet their needs for the conservation of plant genetic resources and to strengthen links to users; (2) to strengthen and contribute to international collaboration in the conservation and use of plant genetic resources, mainly through the encouragement of networks on both a crop and a geographical basis; (3) to develop and promote improved strategies and technologies for the conservation of plant genetic resources; and (4) to provide an information service to advise the world's genetic resources community of both practical and scientific developments in the field (Raymond 1993).

FAO   The Food and Agricultural Organization, headquartered in Rome, has long been involved in genetic resources conservation. As a direct result of the conferences held in the 1960s and other meetings of the FAO Panel of Experts on Plant Exploration and Conservation, a much greater awareness of the importance of crop genetic resources was fostered. The formation of IBPGR was only one outcome of this effort. Another major development was the formation of the FAO Commission on Plant Genetic Resources (CPGR) in 1983. The CPGR is a unique intergovernmental forum, where countries that were donors or users of germplasm, funds, and technology could seek consensus on subjects of global interest. As of April 1993, 117 nations, including the United States, Canada, and Mexico, had become members of the CPGR (Anonymous 1993).

Another FAO effort concerns the International Undertaking on Plant Genetic Resources which was established by a resolution adopted on November 23, 1983, at the Twenty-second Conference of FAO in Rome. The Undertaking was to be a voluntary action by governments to safeguard genetic resources and ensure their "unrestricted availability for the purposes of plant breeding and agricultural development of all countries" (Anonymous 1984). Adherence to the Undertaking was pledged by 107 countries as of April 1993 (Anonymous 1993). At present, neither the United States nor Canada has adhered to the Undertaking; however, as the CPGR oversees the actions of the Undertaking, Canada and the United States participate in the program through membership in the commission.

A significant event affecting worldwide germplasm efforts, including those of the CGIAR centers and FAO, was the United Nations Conference on Environment and Development (UNCED), also referred to as the Earth Summit, which was held in Rio de Janeiro, June 3–14, 1992. Some

180 governments were present at this historic meeting that presented a Convention on Biological Diversity. Agenda 21 (see *Diversity*, vol. 8, no. 3, for text of the conference) describes the various actions and activities proposed for the future. An initial reluctance by the United States to sign the resolution was based on concerns that the biotechnology provisions might negatively impact U.S. businesses. By the end of July 1993, 165 countries had signed, and by September 29 the required 30 countries had ratified the document, thus allowing it to go into effect as of December 29, 1993 (Chasek 1994). The United States has signed the Biodiversity Convention, and it is awaiting ratification by the Senate.

United States National Plant Germplasm System

The United States has one of the most advanced programs for preservation of genetic resources, the National Plant Germplasm System (NPGS). The United States is described as a "germplasm-poor" country (for primary food crops); most of the crops we utilize in our vast agricultural industry have been introduced from other countries. This has necessitated a strong program of introduction, evaluation, utilization, and preservation of genetic resources. Management of the NPGS is through the U.S. Department of Agriculture's Agricultural Research Service (USDA-ARS). Active participants include federal, state, and private-industry scientists working cooperatively to ensure that we have an adequate and safe food supply for future generations. The reader is referred to *Plant Breeding Reviews*, vol. 7, which contains a thorough review of all aspects of NPGS (Janick 1989). Here we will only briefly review the system, with special emphasis on the NSSL.

History, mission, components   The history of formal plant introduction into the United States dates back to at least 1772, when Benjamin Franklin sent a letter to Noble Wimberly Jones referring to a shipment of upland rice from China and seed of the Chinese tallow tree. This practice of obtaining plant material from other countries was quite common, and President John Quincy Adams formalized the practice in 1827 when he requested all consular officials outside the United States to forward rare seeds and plants to Washington for distribution (Hyland 1984). With the establishment of the USDA in 1862, the commissioner of agriculture was specifically directed "to collect, as he may be able, new and valuable seeds and plants; to test, by cultivation, the value of such of them as may require such tests; to propagate such as may be worthy of propagation, and to distribute them among agriculturalists" (Hyland 1984:7). In 1898 the Section of Seed and Plant Introduction was created and the first PIs (Plant Introductions) were collected and numbered; these serve as the beginning of what we now call NPGS. Since 1898 more than 575,000 seed and plant accessions have received PI numbers. Of these, about 425,000 accessions, representing more than 8,250 species, still exist within the NPGS.

The overall mission of the NPGS is to provide "the genetic diversity necessary to improve crop productivity and to reduce genetic vulnerability in future food and agricultural development, not only in the United States but for the entire world. The NPGS acquires, maintains, evaluates, and makes readily accessible to plant scientists a wide range of genetic diversity in the form of seed and clonal germplasm of crops and potential new crops" (White, Shands, and Lovell 1989:18–19).

In order to carry out the mission of the NPGS, the U.S. Department of Agriculture, in cooperation with agricultural experiment stations of several state universities, has established research and preservation centers at numerous locations around the United States (table 11.1). For seed-propagated plants, the majority of the germplasm is held at one of five stations: the four Regional Plant Introduction Stations located in Ames, Iowa; Geneva, New York; Griffin, Georgia; Pullman, Washington; and the National Small Grains Collection located in Aberdeen, Idaho. Genetic stock collections of selected species are held at various Genetic Stock Centers, primarily located on university campuses. Clonally propagated species of fruit and nut crops are held at the National Clonal Germ Plasm Repositories (table 11.1). Each location has specific crops assigned as priority species; however, some species may be assigned to more than one location. Funding for all the locations is provided through the USDA-ARS and from the various university agricultural experiment stations. Private industry has provided additional financial support in the form of special grants and has assisted in the regeneration of seed accessions.

National Seed Storage Laboratory (NSSL)   Nikolai Ivanovich Vavilov, the Russian botanist, established the first gene bank, the All-Union Institute of Plant Industry, in Leningrad (now St. Petersburg) in the 1920s (Plucknett et al. 1987). At the time cold-storage facilities were not available; consequently, accessions had to be regrown frequently to maintain viability. Many of Vavilov's collections were still available in

TABLE 11.1
*Location of and Brief List of Crops at Germplasm Repositories
in the National Plant Germplasm System*

Regional Plant Introduction Stations

Ames, IA	amaranth, artichoke (Jerusalem), asparagus, beet, bentgrass, buckwheat, cantaloupe, carrot, chicory, clover (sweet), collard, coriander, corn, crambe, cucumber, dill, dogwood, endive, gourd, honeydew melon, horseradish, kale, kohlrabi, muskmelon, mustard, ornamentals, parsley, parsnips, pawpaw, pumpkin, rutabaga, spinach, squash, sugarbeet, sunflower, turnip, zucchini
Geneva, NY	artichoke, birdsfoot trefoil, broccoli, brussels-sprouts, cabbage, cauliflower, celery, Chinese cabbage, clover, grasses, ornamentals, legumes (forage), onion, pumpkin, radish, shallot, squash, tomato
Griffin, GA	Bermudagrass, blackeyed pea, castor bean, clover, eggplant, forage legumes, gourds, grasses, guar, kenaf, lespedeza, luffa, mungbean, okra, peanut, pearl millet, Pennisetum, pepper, pigeonpea, pumpkin, Serradella, sesame, sorghum, squash, sweetpotato, vetch, water chestnut, watermelon, wingbean, zoysia grass
Pullman, WA	alfalfa, beans, bluegrass, brome, canarygrass, chickpea, chive, fescue, garlic, grasses, leek, lentil, lettuce, Lupine, milkvetch, onion, orchardgrass, pak choi, pea and pea genetic stocks, ryegrass, safflower, sainfoin, teff, vetch, wheatgrass, wildrye

National Small Grains Collection

Aberdeen, ID	Aegilops, barley and barley genetic stocks, oats, rice, rye, Triticale, wheat

National Clonal Germplasm Repositories

Brownwood, TX	chestnut, hickory, pecan
Corvallis, OR	blackberry, blueberry, boysenberry, cranberry, currant, filbert, gooseberry, hazelnut, hops, ornamentals, mint, pear, raspberry, strawberry
Davis, CA	almond, apricot, cherry, fig, grape, kiwifruit, nectarine, olive, peach, persimmon, pistachio, plum, plumcot, pomegranate, tomato genetic stocks, walnut

Geneva, NY	apple, grape
Hilo, HI	carambola, guava, lychee, macadamia, papaya, passionfruit, Pilis, pineapple, rambutan, tropical plants
Mayaguez, PR	bamboo, banana, Brazilnut, cashew, cassava, cocoa, plantain, sweetpotato, tanier, tropical plants, yam
Miami, FL	avocado, cassava, mango, ornamentals, passion fruit, sugarcane, Tripsacum, tropical plants (note: materials are to be moved to Mayaguez and/or Hilo)
Riverside, CA	date, grapefruit, lemon, lime, orange, tangerine
Sturgeon Bay, WI	potato
Washington, D.C.	dogwood, holly, ornamental plants, magnolia, maple, oak, rhododendron

Genetic Stock and Special Collections

Brookings, SD	native grasses
College Station, TX	cotton, cotton genetic stocks, sorghum genetic stocks
Columbia, MO	wheat genetic stocks
Fargo, ND	flax, wheat (durum) genetic stocks
Lexington, KY	clover
Logan, UT	grasses (forage, range)
Oxford, NC	tobacco (note: to be moved to Griffin)
Raleigh, NC	gamagrass (note: to be moved to Mayaguez)
Salinas, CA	endive, lettuce
Tifton, GA	grasses (wild), pearl millet
Urbana, IL	maize genetic stocks, soybean, soybean genetic stocks

National Seed Storage Laboratory

Fort Collins, CO	apple buds, base collection of all seed crops, genetic stocks (barley, tomato, and wheat)

the 1970s when the capacity for long-term storage was acquired (Plucknett et al. 1987). The NSSL was the first facility designed specifically for long-term seed germplasm preservation and was opened in the fall of 1958 in Fort Collins, Colorado. Previously, seeds were held in medium-term storage at one of the Regional Plant Introduction Stations.

The original NSSL was built from prestressed concrete and comprised about 20,000 square feet of floor space with ten cold-storage vaults having a total area of about 4,500 square feet. The vaults were designed to hold 180,000 samples of one pound each. Seeds were stored in screw-top metal cans (later changed to sealed foil polyethylene laminated bags) that were placed in steel trays on stationary racks. Periodic germination retests (usually at five-year intervals) detected when seeds began to deteriorate and needed to be regrown.

In 1992 a 65,000-square-foot addition to NSSL was dedicated, along with a total renovation of the existing building. The new facility is a state-of-the-art germplasm center containing 20,000 square feet of available storage space, of which 10,000 square feet are reserved for cryogenic storage using LN. Storage capacity for a million accessions is available, along with space for seed receiving, processing, germination, and research. The use of LN as a long-term storage medium for seeds and other plant genetic materials required that special design features be incorporated within the new facility (table 11.2).

Research conducted at NSSL is broadly directed toward developing long-term strategies for preservation of all forms of germplasm of seed and clonally propagated crops (Roos 1992). Heavy emphasis is placed on cryogenic storage in LN. Basic research emphasizes seed and cell biochemistry and physiology as it relates to deterioration.

Seed Preservation Technologies

Before proceeding to discuss seed preservation technologies, it is important to distinguish the two major categories of seeds, based on their storage behavior. "Orthodox" seeds can be readily dried to low moisture content and survive exposure to low (subfreezing) temperatures. This category includes most of our common agricultural plant species and many other wild species, usually (but not always) of temperate climate origin. Orthodox seeds are generally easy to store, have a longevity of several years, decades, or even centuries, and comprise the bulk of species preserved in gene banks. "Recalcitrant" seeds, often

TABLE 11.2
Design Features of Cryopreservation Storage Area

General Design	The cryopreservation storage facility should be composed of three areas: a bulk LN storage area, the accession cryogenic storage (Cryovats) room, and a storage room observation area.
Air-Handling Systems	Cryovat room. The primary air-handling system is in continual operation designed to provide lateral air flow patterns, starting from the top center of the room flowing downward to the lower outside edges of the room. The input air should be 100% outside fresh air. The inlet air temperature can be uncontrolled (i.e., outside ambient air temperature) down to a low limit of approximately 12°C. The volume of air flow should allow 4 room air changes per hour. A secondary (emergency) system should be incorporated to be activated manually with switches and panic buttons or automatically by an oxygen-monitoring system. The fans and ducting for the emergency system should be separate from the primary system. The airflow pattern is longitudinal, flowing perpendicular to the primary airflow patterns. When combined with the primary system, a total of 12–14 room air changes per hour should be achieved. The emergency airflow system should have the same type of temperature control as the primary system. Observation area. This area has its own air-handling system, separate from the cryovat room. Airflow and temperature control are similar to design criteria for general purpose rooms. Temperature should be controlled to approximately 21°C, 100% fresh air intake, with 2–4 air changes per hour. Liquid nitrogen outside storage area. This area is an outside area located next to the cryovat storage area. There should be concrete or metal walls around the area for security. There is no roof. Vents are located around the base of the walls allowing for natural airflow inside the secured area.
Oxygen-Monitoring System	Ambient oxygen monitors should be located throughout the cryovat storage area (minimum of 2) and observation area. The monitors should have two alarm levels, approximately 19% as a first warning and 18.2% as the emergency warning. Each unit has its own oxygen-level display. When an oxygen-monitoring unit goes into 19%-level alarm, the emergency airflow system should be activated and the LN delivery line solenoids put into a shutoff position. If the oxygen monitors go into the second alarm, appropriate personnel should be alerted and emergency-condition procedures enacted.
Manual Activation Stations	There should be manual emergency activation stations in the cryovat area and one observation area. When pressed, they activate the same sequence of events as described above for the oxygen-monitoring system.
Personnel-Monitoring Systems	Voice and video monitoring should be provided for all rooms in the cryostorage area.
Observation Area(s)	The observation area acts as a point of control for the cryovat storage rooms. Monitoring equipment, video displays, air-handling controls, phones, and emergency equipment should be located in this area. The airflow in this room is separated from the cryovat room in such a way that it will provide a safe area if the cryovat room goes into emergency alarm conditions.

termed *desiccation sensitive*, are characterized by their inability to lose water and remain viable under natural conditions. Plants from tropical rain forests, riparian environments, or some temperate tree species often produce recalcitrant seeds. Seeds are usually large and have a longevity under natural conditions of only days, weeks, or a few months, at most. Because of their high moisture content, seeds cannot survive exposure to freezing temperatures. Germplasm of these species is usually held as clonal material and is propagated vegetatively. As a result of very recent research on controlled desiccation, however, gene banks are now beginning to hold seed samples of recalcitrant species.

Seed Longevity

In this chapter seed longevity will be defined as the period of time that a seed retains any spark of life. The capability of many seeds to lose most of their cellular water and not lose their viability is a key step to seed longevity. In the "dry" state, seed longevities may approach years, decades, and in some instances centuries, depending on the species and storage conditions. Even under good storage conditions, seeds will eventually deteriorate: first they lose their "vigor," then the ability to germinate, and finally all viability.

Bewley and Black (1982), Priestley (1986), Roos (1986), and Toole (1986) have recently reviewed the topic of seed longevity. Older reviews by Crocker (1938), Owen (1956), Barton (1961), Harrington (1972), Roberts (1972), and Justice and Bass (1978) all contain excellent information and special insight on this topic. The classic compilation of data on seed longevity by Ewart (1908) is still a primary source of information for many species. Bass (1981) compiled tables of seed longevities for many agricultural and horticultural species stored under various conditions. A summary of reports on seed longevities of a hundred years or longer is presented in table 11.3.

Several reports have claimed extreme longevities for many seeds. The most notorious of these involves the survival of wheat seeds for thousands of years in the tombs of Egyptian kings and princes (table 11.3). These so-called mummy wheats (actually wheat and barley) were part of the supplies placed in the tombs for use by the deceased in the "Land of the Dead." The seeds could be used for food or to plant crops for the years ahead. Reports of such survival in Egyptian tombs have been published in various forms since 1840 (for a discussion, see Barton

TABLE 11.3

Claims of Seed Longevities of 100 or More Years

Species	Longevity	Comments	Reference
Claims lacking direct scientific corroboration			
1. *Lupinus arcticus* S. Wats.	10,000+	[14]C dating of assoc. rodent bones	Porsild, Harrington, and Mulligan, 1967
2. "Mummy" wheat and barley	2,000–3,000+	Egyptian tombs	Barton 1961; Justice and Bass 1978, Bewley and Black 1982
3. *Nelumbo nucifera* Gaertn.	3,075	[14]C dating of assoc. wood canoe	Libby 1954; Wester 1973
4. *Chenopodium album* L.	1,700	Archeological-dated soil samples	Odum 1965
Spergula arvenis L.	1,700		
5. *Nelumbo nucifera* Gaertn.	120–1,080	[14]C dating, geologic, and historical records	Ohga 1923, 1927; Libby 1951; Godwin and Willis 1964
6. Various weed species (36 different species)	110–600+	Archeological-dated	Odum 1965
7. *Canna compacta* Rosc.	530–620	[14]C dating of assoc. materials	Sivori, Nakayama and Cigliano 1968; Lerman and Cigliano 1971
Claims substantiated by direct observation			
8. *Nelumbo nucifera* Gaertn.	237	Museum specimen	Anonymous 1942
9. *Medicago polymorpha* L.	200	From adobe bricks	Spira and Wagner 1983
Hordeum leporinum Link	200	Positive TZ reaction	
Malva parviflora L.	183–200	Positive TZ reaction	
Trifolium sp.	193	Positive TZ reaction	
Chenopodium murale L.	183	Positive TZ reaction	
Chenopodium album L.	143	Positive TZ reaction	
10. *Cassia multijuga* A. Rich.	158	Museum specimen	Becquerel 1934
11. *Albizzia julibrissin* Durazz.	147	Museum specimen	Anonymous 1942
12. Barley and oats	123	Building cornerstone	Aufhammer and Simon 1957
13. *Hyoscyamus niger* L.	116	Museum specimen	Odum 1965
14. *Cassia bicapsularis* L.	115	Museum specimen	Becquerel 1934
15. *Goodia lotifolia* Salisb.	105	Museum specimen	Ewart 1908
Hovea linearis (Sm.) R. Br.	105	Museum specimen	
16. *Malva rotundifolia* L.	100	Buried seed	Kivilaan and Bandurski, 1981
Verbascum blattaria L.	100	Buried seed	
Verbascum thapsus L.	100	Buried seed	
17. *Trifolium pratense* L.	100	Museum specimen	Youngman 1951
Lotus uliginosus (*L. pedunculatus* Cav.)	100	Museum specimen	

Source: Data from Roos 1989.

1961; Priestley 1986; and Toole 1986); however, no documented scientific evidence exists to support the claim that seeds have remained alive for thousands of years (Justice and Bass 1978). The folklore of ancient seeds from Egypt germinating has resulted in other claims for seeds that supposedly were taken from King Tutankhamen's tomb. Seeds of "King Tut" pea and "King Tut" bean are in storage at the NSSL. The origin of these seeds is assuredly from modern plant breeders.

As with the case of survival of seeds buried in tombs, survival of seeds in soil has associated tales of extreme longevity. The oldest claimed longevity was for seeds of the arctic lupine (*Lupinus arcticus* S. Wats.) which were estimated to be more than ten thousand years old (Porsild, Harrington, and Mulligan 1967). The seeds were found in July 1954 during a placer mining operation in the Yukon in lemming burrows deeply buried in permanently frozen soil. The dating of these specimens is based on ^{14}C dating of similar burrows and nests of small rodents found in central Alaska. One such test showed the remains of an arctic ground squirrel to be $14,860 \pm 840$ years old (table 11.3). Unfortunately the dating was not done on the actual seeds, and there is also some question about the seeds themselves as they were not evaluated for several years after their discovery and had changed hands in the interim (Roos 1986).

Viable buried seeds of the sacred or Indian lotus (*Nelumbo nucifera* Gaertn.) have been carbon-dated at anywhere from 120 to over 3,000 years (see Roos 1986). In one study Ohga (1923) dug the seeds out of a layer of peat in a naturally drained lake bed near the village of Pulantian in southern Manchuria. Libby (1951), using ^{14}C analysis, originally dated these seeds at 1040 ± 210 years; however, a later test on these same seeds placed the age at 100 ± 60 years (Godwin and Willis 1964). In a more recent study (Priestley and Posthumus 1982), four seeds from the same site (collected in 1951) were germinated. Two of the seeds were dated at 340 ± 80 years and 430 ± 100 years.

Documented seed longevities   It is not surprising that the oldest scientifically documented seed longevity is that of the Indian lotus (table 11.3). During the period from 1843 to 1855 Robert Brown germinated several seeds from the Hans Sloane collection at the British Museum (Barton 1961). At the time of testing, seeds of *"Nelumbium"* (*Nelumbo*) were 150 years old. In 1926 Ohga (Anonymous 1942) visited London and was unable to germinate an additional twelve seeds from this collection. Ramsbottom (Anonymous 1942), however, succeeded in germinating a single seed, presumably from this same collection, which placed its age at 237 years or possibly as much as 250 years. Other museum specimens having notable seed longevities include *Cassia multijuga* A. Rich., 158 years; *Albizia julibrissin* Durazz., 147 years; as well as several others over 100 years (table 11.3). Aufhammer and Simon (1957) reported germination of seeds of barley and oats after 123 years of storage in a cornerstone of a building in Nuremberg, Germany.

Longevity experiments   Several studies have been initiated in which seeds were buried in the soil and later dug up and tested for viability (see discussion in Crocker 1938, Barton 1961, Priestley 1986, Roos 1986, and Toole 1986). The most famous of these studies was initiated in 1879 by Professor W. J. Beal of the Michigan Agricultural College (Michigan State University) and is still under way. The last report (Kivilaan and Bandurski 1981) showed germination of three species (out of the original twenty) after a hundred years in the soil (table 11.3). Six bottles remain to be excavated and tested over the next sixty years. Other buried seed studies include those of Duvel (1905), Goss (1933, 1939), Kjaer (1940, 1948), Toole and Brown (1946), Lewis (1958, 1973), Madsen (1962), and Egley and Chandler (1978, 1983).

Although buried seed experiments give us some information on seed longevity under natural conditions, controlled environment storage, as in modern gene banks, could extend viability to much longer periods. Three long-term storage experiments are currently being continued at NSSL. The first involves a vegetable-seed longevity study first reported on by James, Bass, and Clark (1964). Seeds of fifteen vegetable species were germinated after storage for fifteen to thirty years in an office in Cheyenne, Wyoming. Retesting some of these same lots twenty-nine years later showed that some viability was still present after fifty to sixty years (Roos and Davidson 1992), the greatest reported longevity for most of these species.

The second experiment, initiated by Went and Munz (1949), concerned storage of ninety-eight species and varieties of California desert plants. The seeds were dried and sealed *in vacuo* and then held at 10 to 20°C. For seventy different kinds of seed, drying *in vacuo* over phosphorus pentoxide for several weeks did not affect germination measurably (49.9 to 43.7%). Germination did not appear to decline over the next five (47.7%), ten (48.05%), and twenty (61.0%) years (Went 1969). This experiment was designed to continue until the year 2307. Recently we were able to obtain the remaining seeds of this experiment from the Rancho Santa Ana Botanical Garden in Claremont, California, where they had been stored unnoticed for several years. These seeds have been transferred to the NSSL and will be tested in the coming year. To our knowledge, they have never been stored under subfreezing conditions.

The third experiment is a continuation of a storage study reported by Rincker in a series of papers (Rincker and Maguire 1979; Rincker 1980, 1981, 1983). He examined the effects of subfreezing storage on forage grass and legume seed germination and forage production. In one study, 291 seed lots representing seven species were stored for up to twenty years at -15°C (Rincker 1981). Seed lots with an originally high germination percentage generally retained a high germination potential. Seed lots with lower initial quality lost germination more quickly. In a second study, 260 lots of nine forage species were placed in subfreezing storage shortly after harvesting (Rincker 1983). After twenty years only three seed lots lost 20% or more in viability. In experiments on forage production following long-term seed storage, the results indicated that forage yields would not be affected if the seeds were of high quality when initially stored (Rincker and Maguire 1979; Rincker 1980). Remnant seed of more than 3,000 seed lots were transferred from Prosser, Washington, to the NSSL in Fort Collins in order to continue this valuable long-term seed storage study.

Recalcitrant seed longevity   The previous discussion concentrated on orthodox seed longevity. Recalcitrant seeds have a rather brief life duration of a few weeks to several months (Roberts 1973a; Chin and Roberts 1980). Harrington (1972) compiled a table of plant species with short-lived seeds, many of which we would call recalcitrant (the term *recalcitrant* had not come into general use at the time). A list of recalcitrant species and those suspected of having recalcitrant seeds has been compiled by King and Roberts (1980). They listed three factors that may contribute to the short longevity of stored recalcitrant seeds: desiccation injury, chilling injury, and problems associated with the storage of high moisture content (mc) seeds such as microbial contamination and germination during storage. A recent model of the physiology of recalcitrant seeds links their longevity in hydrated storage to their rate of germination (Berjak, Farrant, and Pammenter 1989). Because of the relatively short life span of these seeds, they have not been routinely placed in storage at the NSSL or other long-term storage facilities. Of concern is that many plant species that produce recalcitrant seeds are from tropical rain forests or temperate wetlands and are in danger of extinction. Fortunately research to develop storage technologies for recalcitrant seeds is under way at NSSL and elsewhere.

PREDICTING LONGEVITY.   The advent of modern seed gene banks presents the interesting challenge to preserve our seed genetic resources indefinitely, under optimal conditions, as a safeguard for the future of all peoples of the world. The knowledge gleaned from the past will be enhanced and transferred to future generations. Thus it would be of extreme importance if we could predict how long seeds might survive in storage.

Roos (1986) reviewed some of the early attempts to predict seed longevity. Harrington's "rules of thumb" related the effects of seed mc and storage temperature to the rate of seed aging (Harrington 1963). The rules state that for each 1% increase in seed mc (between 5 and 14%), the life span of the seed is halved; for each 5° increase in temperature (between 0 and 50°C), the life span is halved, independent of the seed mc.

A mathematical model of seed deterioration has been proposed (Roberts 1972; Ellis and Roberts 1981; Roberts 1986). The "basic viability equation" can be used to predict germination after any storage period at any temperature and at any seed mc (within reasonable limits). The equation depends on the determination of several constants, including the initial seed quality (specific for each seed lot) and constants that take into account the effects of temperature, seed mc (determined for each species), and species. Once the constants have been determined for a given species and seed lot, predictions can be made as to longevity under any number of different storage conditions. To determine the temperature and seed moisture constants, seeds were aged under various temperatures and seed mc regimes ranging from 3 to 90°C and 6 to 25% mc, respectively. Users are cautioned to be aware of the limits to the viability equation (Roberts 1986). For example, the equation may not be applicable to cold-storage conditions employed in long-term germplasm facilities, such as the NSSL, or to very low seed mc (Ellis, Hong, and Roberts 1989; Vertucci and Roos 1991, 1993a).

The accelerated aging (AA) test (Delouche 1965; Delouche and Baskin 1973) was originally developed to predict the relative storability of seed lots. In this test seeds are placed under high temperatures (40 to 45°C) and high relative humidities (>90%) for short periods of time (2 to 5 days) and then subjected to the standard germination test. Results are compared to unaged seeds tested at the same time.

Although the basic principle of the AA test (and other rapid aging tests), namely, that seeds having low vigor (see below) will not store as well as those having high vigor, is probably valid, it is not necessarily true that seeds stored under gene bank conditions will behave in the same manner. The mechanism for loss in germinability under slow- and rapid-aging conditions is probably not the same. Evidence for this comes from four observations: (1) under high humidity, high-temperature storage conditions, seed mc is elevated to a point where fungi and other microorganisms can grow readily; (2) under slow-aging conditions the percentage of abnormal seedlings increases as the percentage of germination (usually expressed as a percentage of normal seedlings) drops; under rapid-aging conditions this large increase in abnormal seedlings is seldom seen (possibly because the seeds pass quickly through this stage); (3) the frequency of chromosomal aberrations seen in root tips of germinating seeds, which is normally correlated with the percentage of germination (Roberts 1972; Murata, Roos, and Tsuchiya 1981) is lower in seeds subjected to rapid aging (Roberts, Abdalla, and Owen 1967); and (4) the chemical reactions that occur under high mc and low mc are very different (Leopold and Vertucci 1989; Vertucci 1989a; Vertucci and Farrant 1994).

An approach similar to that of AA for predicting which seed lots might deteriorate first in gene banks involves the use of "artificial aging" techniques (Murata, Roos, and Tsuchiya 1980; Roos and Rincker 1982). The significant differences between AA and artificial aging approaches lie in the use of lower seed mc and storage temperatures and longer storage periods. Typically, seed mc, or relative humidity (RH), should be below levels where active fungal growth can occur (<14% mc, ca. 75% RH), storage temperatures are less than 35°C, and storage periods should be on the order of weeks or months. The objective is to achieve a slower deterioration than under AA conditions, which should more closely approximate the deterioration of seeds in long-term cold storage. Moore and Roos (1982) compared different artificial aging tests to evaluate rates of deterioration and analyzed statistically the quantal responses. Artificial aging tests that employed high humidities and high storage temperatures (90% RH, 32°C) resulted in high data heterogeneity and concomitant lack of reliability for predicting seed survival. More moderate aging conditions (70% RH, 21°C) resulted in low data heterogeneity. A computer program was developed to calculate the deterioration rates and apply the statistical analysis (Moore, McSay, and Roos 1983). This program was used by Priestley,

Cullinan, and Wolfe (1985) to determine differences in seed longevity at the species level.

Seed Deterioration

Theories   Several theories have been advanced to explain why seeds deteriorate, even under relatively good storage conditions. These have been discussed previously (Roos 1989) and are summarized here.

Depletion of food reserves   This theory suggests that over time essential substrates become depleted. The major food reserves (storage carbohydrates, proteins, and lipids) may show minor losses over time as a result of respiration. However, examination of nonviable seeds shows that plenty of reserve food remains. Also, we now know that seeds stored at low mc (less than 20%) have essentially no mitochondrial activity (Leopold and Vertucci 1989; Vertucci 1989a). Thus depletion of food reserves by mitochondrial pathways is likely to be minimal. Loss of enzyme cofactors (vitamins and minerals) has not been linked with loss of viability.

Alteration of chemical composition   Although food reserves may not be depleted, they may be altered chemically so that they are not in a usable form. For example, Crocker and Groves (1915) suggested that seed death resulted from a "coagulation of protein." Proteins undergo changes as seen by decreased solubility, partial breakdown, and decreased digestibility. Also, it is well known that lipids undergo oxidation, and an increase in fat acidity has been associated with deterioration. Many other compounds may change quantitatively or qualitatively during storage.

Membrane alteration   The theory that membrane alteration causes deterioration has its basis in the observation that aged seeds have a great propensity to lose cell constituents on imbibition (Matthews and Bradnock 1968). This "leakiness" is attributed to a loss of membrane integrity. Although the exact mechanism causing the membrane alteration is not known, speculation surrounds the role of peroxidation of unsaturated fatty acids (Priestley 1986; Wilson and McDonald 1986). This theory could be considered a corollary to the previous theory except that the chemical changes directly impact membrane permeability. Much of the current research on seed deterioration is concentrated on examining membrane alteration.

Loss of enzyme activity   Duvel (1902), in his Ph.D. dissertation, and later White (1909) proposed that respiration and enzyme activity were important in the preservation of vitality. Many enzymes have been studied in an attempt to correlate seed viability with the amount of enzyme activity. The basic problem with regard to loss of enzyme activity seems to be that of cause and effect. Is loss of enzyme activity the cause of lost viability or is there another cause that also affects loss of enzyme activity?

Genetic damage   Alteration of the cell's DNA, such that it is unable to duplicate and divide, could occur from either chemical mutagenesis or from ionizing radiation. It is a well-established phenomenon that deteriorating seeds produce an abundance of chromosomal aberrations as seen in root tips of germinating seeds (Roberts 1973b; Roos 1982, 1986). Seed viability has been inversely correlated with the frequency of these aberrations. Again, the mechanism for the production of the aberrations is not known. However, evidence has been advanced that the increase in chromosomal aberrations with aging may be a result of the breakdown in the cell's natural DNA repair mechanism(s) (Villiers and Edgcumbe 1975; Yamaguchi, Naito, and Tatara 1978).

Current research approaches   Because the mechanism by which seeds lose their viability with time is unknown, it becomes difficult to develop methods to prevent the deterioration, i.e., conditions that limit all the possible aging reactions. One of the primary problems associated with seed preservation is determining the optimum conditions for storing seeds. Traditionally, optimum protocols have been predicted using equations such as Harrington's "rules of thumb" (Justice and Bass 1978) or the seed viability equation (Roberts 1973a; Ellis and Roberts 1980), which relies on the extrapolation of observations made at very warm and moist conditions.

At NSSL we have used a biophysical approach to the problem of predicting seed longevity under conditions where collection of actual aging data is impractical (Vertucci and Roos 1990, 1993b). We believe that the nature of water binding in seeds and how bound water affects the physiological status of cells (Clegg 1978; Leopold and Vertucci 1989; Vertucci 1989a, 1990; Vertucci and Farrant 1994) influences the nature

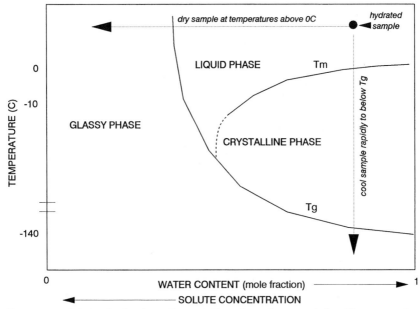

FIGURE 11.1   Schematic of a phase diagram of a water and sugar solution. Tm represents the equilibrium freezing temperature of water, and Tg represents the glass transition temperature of water. The two ways of achieving a glass in a hydrated sample are demonstrated by the dotted lines.

and kinetics of deterioration (Rockland 1969; Karel 1975; Vertucci and Roos 1990, 1993b; Vertucci 1992). This knowledge has served as a foundation for making predictions about optimum seed storage conditions.

Water content and temperature are the most important factors determining seed longevity (Priestley 1986). It has long been established that drying tissues or cooling them preserves their chemical integrity. In effect, these preservation procedures alter the thermodynamic properties of water. Thus to understand seed aging from a thermodynamic standpoint, it is important to understand the properties of water and how these are affected by mc and temperature.

As with all chemical substances, water has three phases: solid (ice), liquid, and vapor. New evidence suggests that water in biological materials can also exist in "glassy" phases (fig. 11.1). A glass appears as a solid, but it is actually a superviscous liquid (Franks 1982). Whereas ice and liquid water represent equilibrium phases, glasses do not. This means that with time aqueous glasses revert to the equilibrium state (i.e., liquid water or ice depending on the temperature). Because they are superviscous, the reversion to the equilibrium state is believed to be slow; thus glasses are considered to be kinetically but not thermodynamically stable (Franks 1982).

Glasses can be formed in two ways: (1) concentrating a solution beyond its level of supersaturation, or (2) cooling a dilute solution to temperatures below the glass transition temperature so rapidly (several thousands of degrees/sec) that ice crystals do not have sufficient time to grow (Franks 1982; Fahy et al. 1984; Burke 1986). Because glasses are believed to be important for preventing freezing injury (Fahy et al. 1984) and desiccation damage (Burke 1986; Williams and Leopold 1989; Koster 1991; Bruni and Leopold 1992) and for promoting longevity in the desiccated state (Vertucci and Roos 1990; Bruni and Leopold 1992; Vertucci and Farrant 1994), glass formation and glass stability (how rapidly a glass reverts to the equilibrium state) is critical in preservation research.

Seed Storage Protocols

Development of optimum seed storage protocols is difficult because of the lack of storage data on long-term survival at near optimum conditions. Nevertheless we have been able to identify the critical factors (mc and temperature) that control the deterioration rate of seeds in storage.

Seed moisture content   We have identified several "types" of water in seed tissues depending on the water content and temperature (fig. 11.2). Water in fully hydrated seeds resembles a dilute solution (type 5). As the seed dries the solution becomes more concentrated (type 4) and finally "vitrifies" (type 3), and then the glassy structure stabilizes (type 2). If seeds are overdried, the glass structure becomes unstable (type 1) (Vertucci 1990, 1994).

Physiological activity corresponds to the different types of water present (Clegg 1978; Rupley, Gratton, and Careri 1983; Priestley 1986; Leopold and Vertucci 1989; Vertucci 1989a; Vertucci et al. 1991; Vertucci and Farrant 1994). Consequently the mechanisms of seed aging also vary with hydration level (fig. 11.3). Deterioration of seeds with water in the glassy phase is a function of how easily substrates are transported to reaction centers, i.e., the fluidity of the system (Burke 1986; Vertucci and Roos 1990; Bruni and Leopold 1992). We believe that deterioration of seeds that are dried so that the glass is disrupted is a result of intermolecular interactions made possible because water has been removed

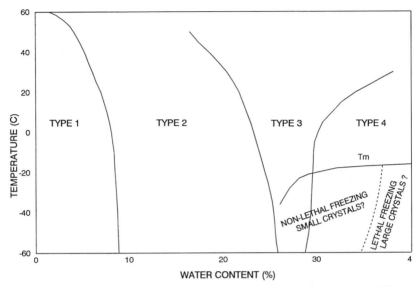

FIGURE 11.2   Phase diagram of the different states of water in soybean seeds. Solid lines were determined from moisture sorption isotherm data or calorimetry (data from Vertucci 1993). Dashed lines are suggested from freezing injury studies (Vertucci 1989b, 1989c).

from the surfaces of macromolecules. This type of damage could be manifested by ionic bonding of charged sites on proteins, membrane phase transitions, or free radical attack on molecules (Rockland 1969; Karel 1975; Vertucci and Roos 1990; Bruni and Leopold 1992; Webb, Hui, and Steponkus 1993; Vertucci and Farrant 1994). Since water protects macromolecules from these interactions, one would expect increased deterioration with the progressive removal of glassy water (Vertucci and Roos 1990).

The idea that different deteriorative reactions occur at different hydration levels has two ramifications: (1) drying tissues progressively does not necessarily enhance seed longevity (as is implied by the longevity equations), and (2) optimizing storage conditions requires determining the water content at which the collective contribution of all deteriorative reactions is minimized. This principle of minimizing damage by optimizing water content can be applied to storage of both recalcitrant and orthodox seeds.

Studies of deterioration in soybean suggest that the optimum mc for storage at 35°C is between 0.05 and 0.08 g H_2O/g dw (5 to 8%) (fig. 11.4).

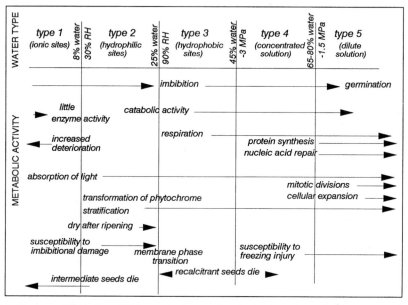

FIGURE 11.3 The effect of different hydration levels on physiological activity in seeds (data from Leopold and Vertucci 1989; Vertucci 1989a; Vertucci and Farrant 1995).

We have demonstrated that optimum moisture levels for storage exist for other orthodox seeds, but that the mc varies according to the chemical composition of the seed and the storage temperature (Vertucci and Roos 1990). However, when expressed in terms of the equilibrium relative humidity (RH) required to obtain the optimum mc, the hydration level for the various seed species is constant and corresponds to the moisture level where the glassy structure is most stable (Vertucci and Roos 1990). This finding suggests that the optimum moisture level for storage of any orthodox seed can be easily achieved by equilibrating to the proper RH and that this practice can be justified by thermodynamic principles.

What is the optimum RH for storage? This is a matter of some debate (Ellis, Hong, and Roberts 1989, 1991; Vertucci and Roos 1990, 1991, 1993a, 1993b; Smith 1992), which will only be resolved by further experimentation. Previously we proposed that optimum moisture level for storage at 25°C could be obtained by equilibrating seeds at 20 to 25% RH (Vertucci and Roos 1990); others have proposed an optimum RH less than this value (Ellis, Hong, and Roberts 1989, 1991; Smith 1992). More recently we have suggested that the optimum RH or water con-

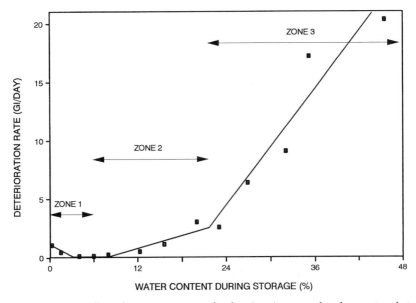

FIGURE 11.4 The effect of water content on the deterioration rate of soybeans stored at 35°C. Deterioration rate is expressed as the rate of decline in the germination index (GI = radicle length x% germination) over the storage period. This was calculated from the regression of GI versus time stored.

tent cannot be considered independently of the storage temperature (Vertucci and Roos 1993b). Theoretically the optimum moisture level for storage of orthodox seeds corresponds to mc and temperature conditions where the glass is most stable. The optimum water content for soybean storage would range from 0.035 g H_2O/g dw at 65°C to 0.11 g/g at -150°C (Vertucci and Roos 1993b). To achieve these mc, seeds dried at 25°C would need to be equilibrated to relative humidities of 6% and 61%, respectively (Vertucci and Roos 1993b). The present protocol adopted at NSSL is to equilibrate all incoming seed samples at 30% RH at 15°C.

Thermodynamic principles relating to the phase behavior of water can be used to predict optimum moisture levels for storage of both orthodox and recalcitrant seeds. The optimum moisture level corresponds to the point at which deterioration resulting from aging reactions, desiccation damage, and freezing injury are minimized. This moisture level is considerably lower for orthodox seeds than for recalcitrant seeds, but for both seed types the optimum mc is predicted to increase with decreasing temperature.

Although the models presented in this paper suggest that optimum moisture levels for storage exist, they do not indicate the actual kinetics of deterioration. The longevity equations provide guidelines for orthodox seeds stored within a stable glassy phase. There are few studies to suggest the longevity of recalcitrant seeds, but it is likely to be related to the stability of the glasses formed. Studies of the kinetics of deterioration and the stability of aqueous glasses are presently a focus of our research.

Storage temperature   Preservation implies slowing the rate of deterioration. Since most reactions require molecular motion, the kinetics of the aging reactions can be altered by manipulating temperature. Generally the lower the temperature, the slower the reactions, and we have postulated that lowering the seed storage temperature to that of LN may provide the ultimate seed storage environment (Stanwood and Bass 1981; Stanwood 1985).

Survival of seeds exposed to temperatures at or near that of LN has been known for nearly 100 years. In experiments of Brown and Escombe (1897), seeds of twelve species, air dried to an mc of 10 to 12%, were slowly cooled in thin glass tubes, immersed in liquid air (-183 to -192°C), and held for 110 hours. After seeds were slowly warmed (over a 50-hour period), there was no apparent effect on germination.

A large number of orthodox seed species has been shown to survive LN exposure (Stanwood and Roos 1979; Stanwood 1980; Stanwood and Bass 1981; Stanwood 1985). Physical damage (splitting, seed-coat cracking) to seeds of some species (garden beans, flax, radish, soybean, and alfalfa) were reported (Stanwood 1985); however, careful control of cooling rate, seed mc, and rewarming rate may overcome this damage (Stanwood 1987; Vertucci 1989b). Seeds cooled at a rate of 1°/hr to -196°C showed only limited damage (Stanwood 1987; Vertucci 1989c).

Recalcitrant seeds   In the above discussion we examined ways of predicting the optimum storage conditions for orthodox seeds. However, these can also be applied to storage of recalcitrant seeds, with the added complication that the moisture level in these seeds must also be adjusted to limit both desiccation damage and freezing injury. For several recalcitrant seeds, this moisture level corresponds to the mc at which water first changes from a concentrated solution to an unstable glass at room temperature (type 3) (fig. 11.2). Successful storage of recalcitrant embryonic axes of *Landolphia kirkii* at -70°C has been achieved using these principles (Vertucci et al. 1991). However, our observations that the optimum mc increases with decreasing temperature suggests that the window of allowable mc gets progressively smaller as the storage temperature is reduced, and eventually freezing injury cannot be prevented without inducing a lethal desiccation damage (fig. 11.2).

The protocols for seed storage described in the preceding paragraphs assume relatively stable or equilibrium conditions. Clearly these protocols are not possible for desiccation-sensitive embryos. An alternative approach is to obtain a glass by rapid cooling to temperatures so low that the glass is stabilized (fig. 11.1). The procedure, known as "vitrification," produces a glass using three principles: (1) concentrate the solution so that the freezing point is depressed and the glass transition temperature increased; (2) cool the sample as fast as possible to minimize the time spent at temperatures below the freezing temperature and above the glass transition temperature; and (3) maintain the sample at a temperature well below the glass transition temperature to prevent the solution from "devitrifying" (Franks 1982; Fahy et al. 1984). Since rapid treatment is essential for success, only small samples can be used. We have successfully applied this procedure to embryonic axes of *Camellia sinensis* (Wesley-Smith et al. 1992) and other species (Wesley-Smith, Berjak, and Pammenter, unpublished; Vertucci and Crane, unpublished) by cooling partially dried axes at about 2,000°C per second. The rapid cooling rates were achieved by injecting axes into subcooled LN, a technique that has recently been adopted to cryopreserve *Drosophila* larvae (Mazur et al. 1992). Freeze-fracture electron microscopy of cells from *C. sinensis* axes treated in this way show no signs of lethal ice formation (Wesley-Smith et al. 1992).

Rescue of Deteriorated Seeds

Unfortunately, gene banks occasionally receive seed in poor condition, with only minimal germination. In many cases this represents valuable germplasm in danger of being lost. Research is under way to rescue this material. The approach usually taken is to provide optimum conditions for germination, including control of microorganism growth, use of seed priming techniques (using osmotic solutions to control imbibition), and providing various nutrients or hormones or both. Alternatively, embryos may be excised and grown in culture to produce plants that can then be transferred to the greenhouse or field, as appropriate. Regeneration of plants from callus and protoplasts may also be a possibility.

Rescue strategies could theoretically target one of several different levels of cellular complexity and organization, ranging from a few remaining living cells in severely deteriorated seeds to isolated embryos to whole seeds. Approaches that exploit the existing morphogenetic program of the embryo should present fewer obstacles than those requiring the induction of morphogenesis in undifferentiated callus tissue (as through embryo- or organogenesis). However, strategies that target the whole seed or embryo might not succeed if only a few cells remain viable and they are severely deteriorated. The optimal approach will thus depend on the seed's state of deterioration. We are currently exploring both tissue culture and embryo germination as options for seed rescue. Work with tissue culture is preliminary but promising. Callus has been successfully initiated from severely deteriorated embryos, and we are attempting to induce somatic embryos from this tissue.

Enhancing the germination of deteriorated seeds, embryos, and embryonic axes is an approach being pursued at NSSL, as well as other laboratories. The seed is an organized group of different organs with a single purpose: to initiate and sustain growth of the embryonic axis. Achievement of this growth requires a sequence of correctly timed and executed biochemical and biophysical processes, all of which are present in incipient form within the healthy dry seed. Deteriorated seed cannot complete this process. It has been assumed that this is because critical elements have deteriorated, for example, substrate pool sizes may be low, enzyme activity may be low, or macromolecular architecture may have been lost.

Strategies that aim to remedy the problem of seed deterioration are as numerous as are the theories of its origin. Intracellular contents leak at a rapid rate from imbibing aged seed (Matthews and Bradnock 1968)—a phenomenon now presumed to result from membrane deterioration. The high concentration of inorganic and organic compounds surrounding the seed provide a rich environment for growth of harmful microbial pathogens. Both surface sterilization and imbibition under sterile conditions have increased the percentage of germination of deteriorated seed in our own as well as in other laboratories (Blackman and Roos 1994). Germination is enhanced if seeds imbibe water slowly, as occurs in a medium with lower water potential (Tilden and West 1985). This treatment is thought to benefit germination by ameliorating the effects of membrane damage.

Another treatment that enhances whole-seed germination is known as priming (Punjabi and Basu 1982; Burgass and Powell 1984; Sur and

Basu 1986; Dell'Aquila and Tritto 1990). During priming, seeds are artificially maintained at water potentials too low for germination to occur but at levels permitting significant metabolic activity (fig. 11.3). Because the sequence of reactions necessary for germination advances, primed seeds germinate faster than nonprimed seeds when transferred to a water potential that supports germination. In some cases the "memory" of priming is retained if the partially hydrated seed is redried following priming. Priming is thought to enhance a deteriorated seed's own repair and reactivation capacity.

Another strategy for rescuing deteriorated embryos or axes is to artificially provide the nutrients that the deteriorated endosperm and scutellum can no longer provide to meet the metabolic and biosynthetic demands of the axis (e.g., Bhattacharyya and Sen-Mandi 1985). The precise mechanism by which endosperm and scutellum contribute to axis growth is obscure. The hydrolysis and transport of macromolecular storage reserves (fat, carbohydrate, and protein) and inorganic reserves such as phytate to the axis requires an exquisitely controlled and complex series of interactions. If any of these are not completed satisfactorily, vigor—and eventually viability—are lost.

We find that sucrose restores the growth of isolated whole embryos (which is normally less than that of whole seeds) to that of the whole seed (Blackman and Roos, unpublished). This effect is observed in embryos from both deteriorated and nondeteriorated seeds. No other nutrient stimulates growth of isolated embryos from deteriorated seed. Thus sucrose appears to be the major limiting nutrient in whole embryos, whether they are excised from fresh or deteriorated seed.

We have tested various hormones for effects on germination of isolated embryos. Auxins have no effect, but kinetin significantly stimulates coleoptile growth and inhibits radicle growth. Gibberellic acid (GA) has two distinct effects. It stimulates coleoptile growth in both fresh and aged isolated embryos and axes. GA also stimulates radicle growth but only in embryos or axes isolated from deteriorated seed. These different effects of GA on aged and fresh seeds may reflect a heretofore unknown aging-induced lesion.

In summary, our ongoing efforts to improve the germination of aged seed in culture have yielded and will continue to yield insights into the physiology of seed deterioration. In addition, NSSL is currently using methods we have developed in a pilot project to save sixty lines of Argentinean maize that have very low field germination and seed numbers (fig. 11.5). Germination rates of embryos isolated from these seeds

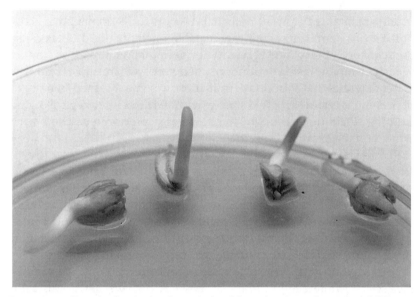

FIGURE 11.5   Growth of excised embryos isolated from deteriorated maize seed (6% germination) on media containing agar, sucrose, fungicide, bactericide, and gibberellic acid. Germination was increased to 79% in culture.

and grown in culture consistently exceed (by 1.3- to 2-fold) germination rates of sterile whole seeds in moistened paper towels (standard laboratory germination test). We are currently refining growth conditions to promote the formation of ear shoots (which has proven somewhat problematic) as the plants mature in the greenhouse.

Preservation of Nonseed Tissues

Although this presentation has mainly concerned preservation of seeds, it should be recognized that important breakthroughs have recently been made on the storage of nonseed germplasm, primarily in LN, on a routine basis. Long-term storage of nonseed germplasm of clonally propagated species is now possible. Preservation of these materials is desirable for providing additional genetic variability (pollen) or for maintaining a specific genetic combination (clonal materials).

Reasons for retaining clonal lines relate to the maintenance of heterozygosity and adapted complexes, inability to set seed easily, long juvenile stages (especially in woody lines), and production of recalcitrant or large seeds (de Langhe 1984). Traditionally clones are maintained in fields or greenhouses and are propagated by grafting, divisions, and so on. This maintenance is costly and clones are subject to loss from disease or disaster. Availability of materials is often seasonal. Tremendous adaptive characteristics associated with survival under different climatic conditions often exist within a species. Growth of all lines at a single location may place some lines under environmental stress and decrease fecundity. Germplasm could be lost.

Advances in micropropagation methodologies now allow for multiplication and maintenance of many species in vitro (Towill 1989). These enhance availability of materials, reduce costs, and minimize disease problems. Other advantages include maintenance of a disease-free state, rapid multiplication rates, ease of rooting (especially for woody materials), reduced maintenance requirements, possibly reduced labor input, and ease of germplasm exchange and shipment both within and between countries. Lines held as in vitro cultures still are susceptible to loss by contamination, equipment failure, or catastrophe. The preservation of species that are vegetatively propagated presents many challenges.

Long-term preservation of these species has been lacking in the germplasm system (Towill and Roos 1989). Long-term preservation (preferably in LN) would provide a convenient backup to active collections, provide safety against loss, and utilize low-cost maintenance systems. Cryopreservation of the clone (shoot tip or bud) is preferred, but is more demanding than that for seed or pollen. Preservation of the latter two preserves most of the gene diversity for the clone, and their storage often is easily accomplished. Details of seed preservation have been described above and are applicable to seeds from vegetatively propagated species (where seeds exist). Seed storage is preferred if the vegetatively propagated species produces desiccation-tolerant seed. If desiccation-sensitive seeds are formed or if seeds are very large, however, pollen preservation is an advantage. Species with desiccation-sensitive seed do not necessarily have desiccation-sensitive pollen.

Pollen

Pollen is not a conventional vehicle for germplasm preservation. A recent review summarizes the current status of pollen preservation

(Hanna and Towill 1995). At present, pollen is stored mainly to facilitate hybridizing materials that flower at different times. However, because of their small size, pollen grains are an efficient way to preserve genes (Towill 1985; Bajaj 1987). A population of pollen grains collected from genetically different individuals contains the nuclear genes within that population. Thus pollen preservation should be viewed as supplemental to preservation of the clone.

Desiccation response   Pollen, as with seed, may be classified (Towill 1985) as either desiccation-tolerant (often bicellular pollen) or desiccation-sensitive (often tricellular pollen). These are arbitrary categories and intermediate types exist, but the classification is useful in understanding what conditions may be necessary to attain long-term preservation. Desiccation-tolerant types store best when dried to a low mc and when held at low temperatures, analogous to orthodox seed. Desiccation-sensitive pollen (for example, pollen from members of the Poaceae) loses viability during drying and usually cannot tolerate low temperatures.

Extensive or systematic studies describing pollen desiccation traits are few. Empirical data derived from exposure to drying and low temperatures have shown that pollen from some fruit crops of temperate and subtropical climates are desiccation-tolerant and survive LN exposure and storage. Examples include apple (Visser 1955), avocado (Sedgley 1981), grape (Parfitt and Almehdi 1983), olive (Parfitt and Almehdi 1984a), some *Prunus* spp. (Parfitt and Almehdi 1984b), pistachio (Vithanange and Alexander 1985), and walnut (Luza and Polito 1988). Information on desiccation characteristics of pollen from many other clonal crops, especially tropical species, are largely anecdotal and in need of more detailed study before long-term preservation is feasible.

Some success has been reported for cooling desiccation-sensitive pollen, notably from maize (Barnabas and Rajki 1976), to very low temperatures, but data are sparse. Careful adjustment of the mc to a specific range appears crucial. Some tricellular pollens, such as from pearl millet (Hanna, Burton, and Monson 1986) and sugarbeet (Hecker, Stanwood, and Soulis 1986), tolerate drying and survive LN exposure. Pollen sugar content appears to be important and provides protection to membranes during desiccation (Hoekstra et al. 1992). Thus metabolic events that determine sugar concentration should affect survival with desiccation and low-temperature exposure.

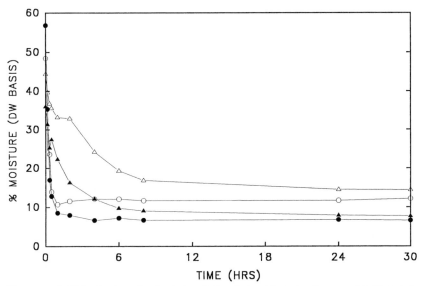

FIGURE 11.6   Dehydration of maize (triangles) and pine (circles) pollen over MgCl$_2$ (solid symbols) and Mg(NO$_3$)$_2$ (open symbols) at -2°C.

Longevities   Surprisingly pollen longevities for most species have not been determined for different moisture and temperature conditions. In general, longevities increase with decreasing temperatures and mc (see Towill 1985 for tables on pollen longevity). As with seeds, there are few reports with quantitative data on longevities at temperatures below about -10°C. Cryopreservation should allow extremely long storage periods, but no long-term storage data are available (Towill 1985). Many reports show no loss in viability in LN over short durations (a few hours to about two years) (Towill 1985; Akihama and Omura 1986). We anticipate that the thermodynamic issues described above for seed water content and storability at different temperatures are also applicable to pollen.

Storage protocols   Handling systems are needed before routine storage is achieved. Some practical aspects have been addressed at NSSL for storage under cryogenic conditions and details depend on species (Connor and Towill 1993). These aspects are similar to issues for handling seed materials. Time of collection of pollen after anthesis is important, and viability of the fresh sample should be determined before desiccation. Drying is usually done over a saturated salt solution (fig. 11.6) to bring the sample to a known mc, but drying under

TABLE 11.4
Germination Values for Fresh, Dry, and Liquid Nitrogen-Stored Pollen

Pollen	% Germination			% Moisture
Type	Fresh	Dry	LN[1]	Dry
Apple	58	17	27	7
Cattail	—[2]	45	43	10
Date	54	59	29	5
Maize	45	49	39	12
Pear	50	47	41	5
Pecan	—[2]	78	61	6
Pine	82	88	84	4
Spruce	87	86	84	9

[1]Time in LN storage for these samples ranges from 24 hours to six months.
[2]Cattail and pecan pollens were dry when collected.

Source: Data from Connor and Towill (1994).

ambient conditions is acceptable if the RH is low; in most cases drying is rapid. The viability of a dried sample should be compared to the fresh sample (table 11.4). Samples should be slowly rehydrated at 95 to 100% RH to moisture levels greater than 25% before testing viability. There are numerous viability tests, but in vitro germination and fluorescein diacetate staining are most common and useful for storage purposes (Towill 1985). For long-term preservation, cryogenic storage is preferred, and exposure to LN or to the vapor phase over LN is by direct immersion. Warming from LN is usually rapid. Cryoampoules are useful storage containers, and samples that are retrieved from storage can be recooled to low temperatures with no loss of viability, providing care is taken not to alter the pollen mc.

Vegetative Propagules

Short-term preservation   Plant parts such as cuttings, budwood, root divisions, and tubers from some species can be stored for about one year in near-freezing temperatures (1 to 5°C) and are used for propagation and distribution. The appropriate storage time varies considerably depending on species. Propagules from chilling-sensitive species require storage at warmer temperatures (10 to 15°C). Most propagules are available only at certain times of the year.

In vitro storage of plants is conducted either under normal growth conditions (such as 20 to 25°C under light), reduced temperatures (-3 to 12°C, with or without light), or on a growth-restrictive nutrient medium at ambient or reduced temperatures. Storage longevities may be weeks, months, and in some cases years before subculture is needed (Aitken-Christie and Singh 1987; Withers 1991). Examples of storage periods include one year for kiwifruit (Monette 1987), sixteen months for garlic (El-Gizawy and Ford-LLoyd 1987), eighteen months for pear (Wanas, Callow, and Withers 1986), and four years for apple (Druarte 1985). Information is scant describing whether a single protocol can be used for diverse genotypes within a species or whether regenerants after storage are normal and easily acclimatized.

Potential problems with the application of in vitro plant preservation include difficulties in culture establishment, micropropagation, rooting, and acclimatization. Many of the details for these stages are cultivar specific. Explants from mature woody plants often are more difficult to establish in culture than are those from juvenile plants. Diverse genotypes may not survive in the same physical and chemical environment. Growth abnormalities may occur. There is an uncertainty about the occurrence of somaclonal variants. Adequate characterization of each in vitro line (e.g., electrophoretic profiles: isozymes, restriction fragment length polymorphisms) must be established and lines periodically examined to avoid identification errors.

Long-term preservation   Cryopreservation of meristem tips, shoot tips, or buds is potentially useful for clonal base collections since it provides a lower-cost and space-saving method for storage. Two general strategies are used. The first utilizes the ability of some species to acclimate to low temperature whereas the second utilizes manipulations (pretreatments, desiccation, and the addition of cryoprotectants) to enhance the survival of species that are less or not at all cold hardy.

DORMANT VEGETATIVE BUDS. The extreme cold hardiness that develops in some woody species is well documented. Several investigations have shown that bark and dormant vegetative buds, when cooled under controlled conditions, survive LN exposure (Sakai and Nishiyama 1978; Tyler, Stushnoff, and Gusta 1988). Reducing the mc in twigs before cooling may enhance survival in tender lines (Tyler and Stushnoff 1988). This strategy should be useful for preserving the germplasm of apple, pear, blueberry, *Prunus* spp., gooseberries, and currants, as well as others. Treated buds are grafted to root stocks to generate a shoot. Alternatively,

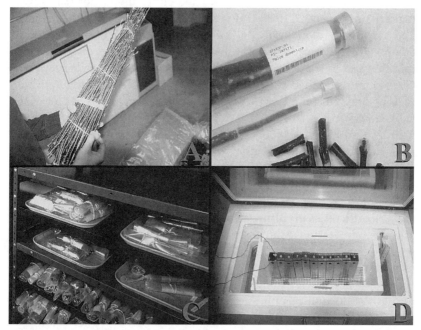

FIGURE 11.7 Preparing winter-hardy twigs of apple for cryopreservation: (A) twigs are harvested in midwinter and shipped to the NSSL; (B) upon receipt twigs are cut into one- or two-bud segments; (C) bud segments are desiccated to 25 to 30% mc at -5°C; (D) buds are packaged in half- or one-inch-diameter polyolefin tubing (see B) for storage, and the temperature is gradually lowered (1°/hr) to -30°C before placing samples in long-term storage over LN (-160°C).

shoot tips can be dissected from LN-treated buds and cultured to produce a shoot. Any method must take into account that genotypes within a species differ in hardiness.

Apple germplasm is currently being cryopreserved at the NSSL using the dormant vegetative-bud technique (Forsline et al. 1993). Winter-hardy twigs are cut into one- or two-bud sections (fig. 11.7), partially desiccated, and gradually cooled (1°/hr) to -30°C before being directly placed into the vapor phase above LN. To date, 313 lines have been examined, and survival after LN has occurred in 289 of them. Buds from cold-hardy genotypes survived in higher percentages than from less hardy lines. Fire-blight-sensitive lines are a priority and are being processed. Apple, therefore, is the first species at NSSL for which clones are being placed into the base collection.

SHOOT TIPS Species that acclimate to cold to a lesser extent or those that do not acclimate at all require additional manipulations before cryopreservation is successful. Shoot tips from field, greenhouse, growth chamber, or in vitro plants are excised for treatment. In some cases cold treatment of the stock plant is beneficial (Reed 1988). After excision the sequence of steps involves preculture (a few hours to two days) in growth medium or with low levels of cryoprotectants; controlled exposure to a defined concentration of cryoprotectant; ice nucleation of the sample just below the freezing point of the solution; controlled cooling (about 0.25 to 1°C/min) to -35 to -40°C followed by transfer directly into LN; rapid warming; and post-thaw treatments before culturing to determine viability. Details for several of these steps have been reviewed (Sakai 1986; Withers 1987; Towill 1990a). Shoot tips from some herbaceous species (for example, carnation, cassava, chickpea, *Digitalis*, pea, peanut, potato, strawberry) and from woody species (apple) can survive LN treatment.

Within the past four years, modifications of the general method have given high levels of survival after LN exposure in shoot tips from several species. These modifications use conditions that probably vitrify cells during the cooling stage (fig. 11.1). The use of cryoprotectant solutions to stabilize the glass is important for successful cryopreservation of shoot tips. Vitrification could eliminate certain injury causing events during cooling/thawing with conventional procedures and may increase the allowable size of the propagule being treated (Fahy, Levy, and Ali 1987). High concentrations of cryoprotectants usually are necessary, and toxicity may be a problem.

Vitrification involves loading the cells with cryoprotectants and then desiccating the system by applying high concentrations of osmotica (Steponkus, Langis, and Fujikawa 1992). With this procedure we have observed high levels of viability in several species such as mint, potato, papaya, *Arachis* sp., apple, *Prunus* spp., and sweet potato. Whether loading is necessary is a concern, and survival with LN often occurs with desiccation alone. The kinetics of loading and dehydration are important and probably differ among species but are not easily measured in small systems, such as cells or shoot tips. A modification of the usual vitrification technique involves coating the excised shoot tip with alginate before any treatment. The alginate capsules are either slowly or quickly exposed to elevated sucrose levels and then desiccated. Capsules are immersed in LN. Survival has occurred with shoot tips from pear (Dereuddre et al. 1990), potato (Fabre and Dereuddre 1990), *Dianthus* spp. (Tannoury et al. 1991), grape (Plessis, Leddet, and Dereuddre 1991), and sugarcane (Paulet, Engelmann, and Glaszmann 1993).

Vitrification is receiving more emphasis than conventional two-step cooling methods, and has thus far been applicable to tender materials such as potato and grape. At NSSL we have obtained survival after LN-exposure of shoot tips from in vitro plants of several *Mentha* species, five potato cultivars, papaya, sweet potato, apple, two *Prunus* species, *Arachis glabrata*, *Phytolacca dodecandra*, and one grape cultivar, and of somatic embryos from cotton and walnut (for methods, see Towill 1990b, 1994; Towill and Jarret 1992). Viability did not decrease for materials held in LN for one year. This, in accordance with data appearing in the literature for protoplasts (Langis and Steponkus 1990), cell suspensions (Langis et al. 1989), and somatic embryos (Uragami et al. 1989), demonstrates the method's usefulness.

Many questions remain. For example, there is the question of whether a single procedure works well for a diverse array of genotypes, and whether plant regenerants are completely normal after treatment. One concern is the maintenance of organization within the shoot tip after treatment and regeneration of the shoot without (or with minimal) callus formation or adventitious bud development (Towill 1988). Direct development from meristem cells to a shoot decreases the chance for somaclonal variation.

The genetic stability of materials exposed to cryogenic protocols has been questioned. One issue is whether extremely long-term storage exposes samples to background levels of radiation that could produce genetic change or lethality since DNA repair processes are not functioning at low temperatures (Lyon, Whittingham, and Glenister 1977; Ashwood-Smith and Friedmann 1979; Glenister, Whittingham, and Lyon 1984). Few data in higher-plant systems provide answers. Projections from studies with animal cells exposed to high levels of radiation while cryogenically stored suggest that lethality would occur only after several thousand years with background radiation levels. Storage clearly is not indefinite at low temperatures, but results suggest that a few centuries of storage is realistic.

A related issue for hydrated materials that are processed with cryoprotectants is whether treatments occurring during the cryogenic protocol are mutagenic (Ashwood-Smith and Grant 1977). No clear answer has emerged (Calcott and Gargett 1981; Ashwood-Smith 1985), and some studies suggest that mutation rates are not increased. Freeze-drying is mutagenic to microbes, but mutagenicity is usually attributed to events during the drying phase and not to cooling and thawing per se (Ashwood-Smith and Grant 1976). No comparable studies have been done with cells or tissues from higher organisms.

Literature Cited

Aitken-Christie, J. and A. P. Singh. 1987. Cold storage of tissue cultures. In J. M. Bonga and D. J. Durzan, eds., *Cell and Tissue Culture in Forestry*, vol. 2, pp. 285–304. Dordrecht: Martinus Nijhoff.

Akihama, T. and M. Omura. 1986. Preservation of fruit tree pollen. In Y. P. S. Bajaj, ed., *Biotechnology in Agriculture and Forestry. Trees*, vol. 1, pp. 101–112. Berlin: Springer-Verlag.

Anonymous. 1942. Recent work on germination. *Nature* (London) 149:658–659.

——. 1984. FAO calls for international "Undertaking" on plant genetic resources. *Diversity*, no. 6 (March–April): 16.

——. 1993. Spirit of cooperation overshadows disharmonies as FAO Commission marks 10th anniversary. *Diversity* 9 (1–2): 4–6.

Ashwood-Smith, M. J. 1985. Genetic damage is not produced by normal cryopreservation procedures involving either glycerol or DMSO: A cautionary note however on the possible effects of DMSO. *Cryobiology* 22:427–433.

Ashwood-Smith, M. J. and G. B. Friedmann. 1979. Lethal and chromosomal effects of freezing, thawing, storage time, and x-irradiation on mammalian cells preserved at -196°C in dimethylsulfoxide. *Cryobiology* 16:132–140.

Ashwood-Smith, M. J. and E. Grant. 1976. Mutation induction in bacteria by freeze-drying. *Cryobiology* 13:206–213.

——. 1977. Genetic stability in cellular systems stored in the frozen state. In K. Elliot and J. Whelan, eds., *The Freezing of Mammalian Embryos*, pp. 251–272. Amsterdam: Elsevier.

Aufhammer, G. and U. Simon. 1957. Die Samen landwirtschaftlicher Kulturpflanzen im Grundstein des ehemaligen Nurnberger Stadttheaters und ihre Keimfahigkeit. *Z. Acker. Pflanzenbau* 103:454–472.

Bajaj, Y. P. S. 1987. Cryopreservation of pollen and pollen embryos, and the establishment of pollen banks. *Internat. Rev. Cytology* 107:397–420.

Barnabas, R. and E. Rajki. 1976. Storage of maize (*Zea mays* L.) pollen at -196°C in liquid nitrogen. *Euphytica* 25:747–752.

Barton, L. V. 1961. Seed preservation and longevity. New York: Interscience.

Bass, L. N. 1981. Storage conditions for maintaining seed quality. In E. E. Finney, Jr., ed., *Handbook of Transportation and Marketing in Agriculture*. Vol. 2: *Field Crops*, pp. 239–321. Boca Raton: CRC Press.

Becquerel, P. 1934. La longévité des graines macrobiotiques. *C. R. Acad. Sci.* 199:1662–1664.

Bennett, E., ed. 1968. *FAO/IBP Technical Conference on the Exploration, Utilization, and Conservation of Plant Genetic Resources*. Rome: FAO.

Berjak, P., J. M. Farrant, and N. W. Pammenter. 1989. The basis of recalcitrant seed behaviour: Cell biology of the homoiohydrous seed condition. In R. B.

Taylorson, ed., *Recent Advances in the Development and Germination of Seeds*, pp. 89–108. New York: Plenum.

Bewley, J. D. and M. Black. 1982. *Physiology and Biochemistry of Seeds in Relation to Germination*, vol. 2, pp. 1–59. New York: Springer-Verlag.

Bhattacharyya, S. and S. Sen-Mandi. 1985. Studies into causes of non-germination of aged wheat seeds. *Ann. Bot.* 56:475–479.

Blackman, S. A. and E. E. Roos. 1994. Culture of isolated embryos of deteriorated maize seed: A strategy for rescuing germplasm. *HortScience* 29:494.

Brown, H. T. and F. Escombe. 1897. Note on the influence of very low temperatures on the germinative power of seeds. *Proc. Roy. Soc. London* 62:160–165.

Bruni, F. B. and A. C. Leopold. 1992. Cytoplasmic glass formation in maize embryos. *Seed Sci. Res.* 2:251–253.

Burgass, R. W. and A. A. Powell. 1984. Evidence for repair processes in the invigoration of seeds by hydration. *Ann. Bot.* 53:753–757.

Burke, M. J. 1986. The glassy state and survival of anhydrous biological systems. In A. C. Leopold, ed., *Membranes, Metabolism, and Dry Organisms*, pp. 358–363. Ithaca: Cornell University Press.

Calcott, P. H. and A. M. Gargett. 1981. Mutagenicity of freezing and thawing. *FEMS Microbiol. Lett.* 10:151–155.

Chasek, P. 1994. Despite swift ratification, Biodiversity Convention off to a halting start. *Diversity* 9 (4); 10 (1): 15–17.

Chin, H. F. and E. H. Roberts, eds. 1980. *Recalcitrant Crop Seeds*. Kuala Lumpur: Tropical Press.

Clegg, J. S. 1978. Hydration-dependent metabolic transitions and the state of water in *Artemia* cysts. In J. H. Crowe and J. S. Clegg, eds., *Dry Biological Systems*, pp. 117–154. New York: Academic Press.

Connor, K. F. and L. E. Towill. 1993. Pollen-handling protocol and hydration/dehydration characteristics of pollen for application to long-term storage. *Euphytica* 68:77–84.

Crocker, W. 1938. Life-span of seeds. *Bot. Rev.* 4:235–274.

Crocker, W. and J. F. Groves. 1915. A method for prophesying the life duration of seeds. *Proc. Nat. Acad. Sci.* 1:152–155.

de Langhe, E. A. L. 1984. The role of in vitro techniques in germplasm conservation. In J. H. W. Holden and J. T. Williams, eds., *Crop Genetic Resources: Conservation and Evaluation*, pp. 131–137. London: Allen and Unwin.

Dell'Aquila, A. and V. Tritto. 1990. Aging and osmotic priming in wheat seeds: Effects upon certain components of seed quality. *Ann Bot.* 65:21–26.

Delouche, J. C. 1965. An accelerated aging technique for predicting relative storability of crimson clover and tall fescue seed lots. *Agron. Abstr.*, p. 40.

Delouche, J. C. and C. C. Baskin. 1973. Accelerated aging techniques for predicting the relative storability of seed lots. *Seed Sci. Technol.* 1:427–452.

Dereuddre J, C. Scottez, Y. Arnaud, and M. Duron. 1990. Resistance of alginate-coated axillary shoot tips of pear tree (*Pyrus communis* L. cv Beurre Hardy) in vitro plantlets to dehydration and subsequent freezing in liquid nitrogen: Effects of previous cold hardening. *C. R. Acad. Sci. Paris* 310:317–323.

Druarte, Ph. 1985. In vitro germplasm preservation technique for fruit trees. In A. Schafer-Menuhr, ed., *In Vitro Techniques: Propagation and Long-Term Storage*, pp. 167–171. Boston: Martinus Nijhoff/Dr. W. Junk.

Duvel, J. W. T. 1902. The vitality and germination of seeds. Ph.D. diss., University of Michigan, Ann Arbor, 96 pp.

——. 1905. The viability of buried seeds. *USDA Bull.* 83. U.S. Government Printing Office, Washington, D.C., 20 pp.

Egley, G. H. and J. M. Chandler. 1978. Germination and viability of weed seeds after 2.5 years in a 50-year buried seed study. *Weed Sci.* 26:230–239.

——. 1983. Longevity of weed seeds after 5.5 years in the Stoneville 50-year buried-seed study. *Weed Sci.* 31:264–270.

El-Gizawy, A. M. and B. V. Ford-Lloyd. 1987. An in vitro method for the conservation and storage of garlic (*Allium sativum*) germplasm. *Plant Cell, Tissue, and Organ Culture* 9:147–150.

Ellis, R. H., T. E. Hong, and E. H. Roberts. 1989. A comparison of the low-moisture-content limit to the logarithmic relation between seed moisture longevity in twelve species. *Ann. Bot.* 63:601–611.]

——. 1991. Seed moisture content, storage, viability and vigour (correspondence). *Seed Sci. Res.* 1:275–279.

Ellis, R. H. and E. H. Roberts. 1980. Improved equations for the prediction of seed longevity. *Ann Bot.* 45:13–30.

——. 1981. The quantification of aging and survival in orthodox seeds. *Seed Sci. Technol.* 9:373–409.

Ewart, A. J. 1908. On the longevity of seeds. *Proc. Roy. Soc. Victoria* 21:1–210.

Fabre, J. and J. Dereuddre. 1990. Encapsulation-dehydration: A new approach to cryopreservation of *Solanum* shoot tips. *Cryo-letters* 11:413–426.

Fahy, G. M., D. I. Levy, and S. E. Ali. 1987. Some emerging principles underlying the physical properties, biological actions, and utility of vitrification solutions. *Cryobiology* 24:196–213.

Fahy, G. M., D. R. MacFarlane, C. A. Angell, and H. T. Meryman. 1984. Vitrification as an approach to cryopreservation. *Cryobiology* 21:407–426.

FAO/IBPGR. 1992. Report of the expert consultation on gene bank standards. Rome: FAO/IBPGR. 26 pp.

Forsline, P. L., C. Stushnoff, L. E. Towill, J. W. Waddell, and W. F. Lamboy. 1993. Pilot project to cryopreserve dormant apple (*Malus* sp.) buds. *HortScience* 28:478.

Frankel, O. H. 1970. Preface. In O. H. Frankel and E. Bennett, eds., *Genetic Resources in Plants—Their Exploration and Conservation*, IBP Handbk. No. 11, pp. 1–4. Oxford: Blackwell.

Franks, F. 1982. The properties of aqueous solutions at subzero temperatures. In F. Franks, ed., *Water: A Comprehensive Treatise*, pp. 215–338. New York: Plenum.

Glenister, P. H., D. G. Whittingham, and M. F. Lyon. 1984. Further studies on the effect of radiation during the storage of frozen 8-cell mouse embryos at -196°C. *J. Reprod. Fert.* 70:229–234.

Godwin, H. and E. H. Willis. 1964. The viability of lotus seeds (*Nelumbium nucifera* Gaertn.). *New Phytol.* 63:410–412.

Goss, W. L. 1933. Buried seed experiment. *Bull. Calif. Dep. Agric.* 22:302–304.

——. 1939. Germination of buried weed seeds. *Bull. Calif. Dep. Agric.* 28:132–135.

Hanna, W. W., G. W. Burton, and W. G. Monson. 1986. Long-term storage of pearl millet pollen. *J. Hered.* 77:361–362.

Hanna, W. W. and L. E. Towill. 1995. Long-term pollen preservation. *Plant Breed. Rev.* 13:179–207.

Harrington, J. F. 1963. Practical advice and instructions on seed storage. *Proc. Intern. Seed Test. Assoc.* 28:989–994.

——. 1972. Seed storage and longevity. In T. T. Kozlowski, ed., *Seed Biology*, vol. 3, pp. 145–245. New York: Academic Press.

Hecker, R. J., P. C. Stanwood, and C. A. Soulis. 1986. Storage of sugarbeet pollen. *Euphytica* 35:777–783.

Hoekstra, F. A., J. H. Crowe, L. M. Crowe, T. Van Roekel, and E. Vermeer. 1992. Do phospholipids and sucrose determine membrane phase transitions in dehydrating pollen? *Plant Cell Environ.* 15:601–606.

Hyland, H. L. 1984. History of plant introduction in the United States. In C. W. Yeatman, D. Kafton, and G. Wilkes, eds., *Plant Genetic Resources: A Conservation Imperative*, AAAS selected symposium 87, pp. 5–14. Bolder: Westview.

International Board for Plant Genetic Resources. 1993. *Annual Report for 1992.* Rome: IBPGR. 92 pp.

James, E., L. N. Bass, and D. C. Clark. 1964. Longevity of vegetable seeds stored 15 to 30 years at Cheyenne, Wyoming. *Proc. Am. Soc. Hort. Sci.* 84:527–534.

Janick, J., ed. 1989. *The National Plant Germplasm System of the United States, Pl. Breed. Rev.*, vol. 7. Portland: Timber Press.

Justice, O. L. and L. N. Bass. 1978. Principles and Practices of Seed Storage. USDA Handb. 506. Washington: U.S. Government Printing Office.

Karel, M. 1975. Physical-chemical modification of the state of water in foods: A speculative survey. In R. B. Duckworth, ed., *Water Relations in Foods*, pp. 639–656. New York: Academic Press.

King, M. W. and E. H. Roberts. 1980. Maintenance of recalcitrant seeds in storage. In H. F. Chin and E. H. Roberts, eds., *Recalcitrant Crop Seeds*, pp. 53–89. Kuala Lumpur: Tropical Press.

Kivilaan, A. and R. S. Bandurski. 1981. The one hundred-year period for Dr. Beal's seed viability experiment. *Am. J. Bot.* 68:1290–1292.

Kjaer, A. 1940. Germination of buried and dry stored seeds. Vol. 1: 1934–1939. *Proc. Intern. Seed Test. Assoc.* 12:167–190.

——. 1948. Germination of buried and dry stored seeds. Vol. 2: 1934–1944. *Proc. Intern. Seed Test. Assoc.* 14:19–26.

Koster, K. L. 1991. Glass formation and desiccation tolerance in seeds. *Plant Physiol.* 96: 302–304.

Langis, R., B. Schnabel, E. D. Earle, and P. L. Steponkus. 1989. Cryopreservation of *Brassica campestris* L. cell suspensions by vitrification. *Cryo-Letters* 10:421–428.

Langis, R. and P. L. Steponkus. 1990. Cryopreservation of rye protoplasts by vitrification. *Plant Physiol.* 92:666–671.

Leopold, A. C. and C. W. Vertucci. 1989. Moisture as a regulator of physiological reaction in seeds. In P. C. Stanwood and M. B. McDonald, eds., *Seed Moisture*, pp. 51–67. Madison, Wis.: Crop Sci. Soc. Amer. Spec. Pub. 14.

Lerman, J. C. and E. M. Cigliano. 1971. New carbon-14 evidence for 600 year old *Canna compacta* seed. *Nature* (London) 232:568–570.

Lewis, J. 1958. Longevity of crop and weed seeds. Vol. 1: First interim report. *Proc. Intern. Seed Test. Assoc.* 23:340–354.

——. 1973. Longevity of crop and weed seeds: Survival after 20 years in soil. *Weed Res.* 13:179–191.

Libby, W. F. 1951. Radiocarbon dates, II. *Science* 114:291–297.

——. 1954. Chicago radiocarbon dates, IV. *Science* 119:135–141.

Luza, J. G. and V. S. Polito. 1988. Cryopreservation of English walnut (*Juglans regia* L.) pollen. *Euphytica* 37:141–148.

Lyon, M. F., D. G. Whittingham, and P. Glenister. 1977. Long-term storage of frozen mouse embryos under increased background irradiation. In K. Elliot and J. Whelan, eds., *The Freezing of Mammalian Embryos*, pp. 273–281. Amsterdam: Elsevier.

Madsen, S. B. 1962. Germination of buried and dry stored seeds. Vol. 3: 1934–1960. *Proc. Intern. Seed Test. Assoc.* 27:920–928.

Matthews, S. and W. T. Bradnock. 1968. Relationship between seed exudation and field emergence in peas and French beans. *Hortic. Res.* 8:89–93.

Mazur, P., K. W. Cole, J. W. Hall, P. D. Schreuders, and A. P. Mahowald. 1992. Cryobiological preservation of *Drosophila* embryos. *Science* 258:1932–1935.

Monette, P. L. 1987. Organogenesis and plantlet regeneration following in vitro cold storage of kiwifruit shoot tip cultures. *Sci. Hort.* 31:101–106.

Moore, F. D., III, A. E. McSay, and E. E. Roos. 1983. Probit analysis: A computer program for evaluation of seed germination and viability loss rate. *Colorado State Univ. Exp. Stn. Tech. Bull.* 147, 7 pp.

Moore, F. D. and E. E. Roos. 1982. Determining differences in viability loss rates during seed storage. *Seed Sci. Technol.* 10:283–300.

Murata, M., E. E. Roos, and T. Tsuchiya. 1980. Mitotic delay in root tips of peas induced by artificial seed aging. *Bot. Gaz.* 141:19–23.

——. 1981. Chromosome damage induced by artificial seed aging in barley. Vol. 1: Germinability and frequency of aberrant anaphases at first mitosis. *Can. J. Genet. Cytol.* 23:267–280.

Odum, S. 1965. Germination of ancient seeds: Floristical observations and experiments with archaeological dated soil samples. *Dan. Bot. Arkiv* 24(2):1–70.

Ohga, I. 1923. On the longevity of seeds of *Nelumbo nucifera*. *Bot. Mag.* (Tokyo) 37:87–95.

———. 1927. On the age of the ancient fruit of the Indian lotus which is kept in the peat bed in South Manchuria. *Bot. Mag.* (Tokyo) 41:1–6.

Owen, E. B. 1956. The storage of seeds for maintenance of viability. *Bull.* 43, Commonwealth Bureau of Pastures and Field Crops. Commonwealth Agricultural Bureaux, Farnham Royal, Bucks, England. 81 pp.

Parfitt, D. E. and A. A. Almehdi. 1983. Cryogenic storage of grape pollen. *Am. J. Enol. Vitic.* 34:227–228.

———. 1984a. Cryogenic storage of olive pollen. *Fruits Varieties J.* 38:14–16.

———. 1984b. Liquid nitrogen storage of pollen from five cultivated *Prunus* species. *HortScience* 19:69–70.

Paulet, F., F. Engelmann, and J. C. Glaszmann. 1993. Cryopreservation of apices of in vitro plantlets of sugarcane (*Saccharum sp.* hybrids) using encapsulation/dehydration. *Plant Cell Rep.* 12:525–529.

Plessis, P., C. Leddet, and J. Dereuddre. 1991. Resistance to dehydration and to freezing in liquid nitrogen of alginate-coated shoot tips of grape vine (*Vitis vinifera* L. cv Chardonnay). *C. R. Acad. Sci. Paris* 313:373–380.

Plucknett, D. L., N. J. H. Smith, J. T. Williams, and N. M. Anishetty. 1987. *Gene Banks and the World's Food.* Princeton, N.J.: Princeton University Press.

Porsild, A. E., C. R. Harrington, and G. A. Mulligan. 1967. *Lupinus arcticus* Wats. grown from seeds of Pleistocene age. *Science* 158:113–114.

Priestley, D. A. 1986. *Seed Aging: Implications for Seed Storage and Persistence in the Soil.* Ithaca: Cornell University Press.

Priestley, D. A., V. I. Cullinan, and J. Wolfe. 1985. Differences in seed longevity at the species level. *Plant Cell Environ.* 8:577–562.

Priestley, D. A. and M. A. Posthumus. 1982. Extreme longevity of lotus seeds from Pulantian. *Nature* (London) 299:148–149.

Punjabi, B. and R. N. Basu. 1982. Control of age- and radiation-induced seed deterioration in lettuce (*Lactuca sativa* L.) by hydration-dehydration treatments. *Proc. Indian Natl. Sci. Acad.*, pt. B., *Biol. Sci.* 48:242–250.

Raymond, R. D. 1993. Conserving nature's biodiversity: The role of the International Plant Genetic Resources Institute. *Diversity* 9 (3): 17–21.

Reed, B. M. 1988. Cold acclimation as a method to improve survival of cryo-preserved *Rubus* meristems. *Cryo-Letters* 9:166–171.

Rincker, C. M. 1980. Effect of long-term subfreezing storage of seed on legume forage production. *Crop Sci.* 20:574–577.

———. 1981. Long-term subfreezing storage of forage crop seeds. *Crop Sci.* 21:424–427.

———. 1983. Germination of forage crop seeds after 20 years of subfreezing storage. *Crop Sci.* 23:229–231.

Rincker, C. M. and J. D. Maguire. 1979. Effect of seed storage on germination and forage production of seven grass cultivars. *Crop Sci.* 19:857–860.

Roberts, E. H., ed. 1972. *Viability of Seeds.* London: Chapman and Hall.

———. 1973a. Predicting the storage life of seeds. *Seed Sci. Technol.* 1:499–514.

———. 1973b. Loss of seed viability: Chromosomal and genetical aspects. *Seed Sci. Technol.* 1:515–527.

———. 1986. Quantifying seed deterioration. In M. B. McDonald, Jr., and C. J. Nelson, eds., *Physiology of Seed Deterioration*, pp. 101–123. Madison, Wis.: Crop Science Society of America Spec. Pub. 11.

Roberts, E. H., F. H. Abdalla, and R. J. Owen. 1967. Nuclear damage and the aging of seeds with a model for seed survival curves. *Symp. Soc. Exp. Biol.* 21:65–100.

Rockland, L. B. 1969. Water activity and storage stability. *Food Technol.* 23: 1241–1251.

Roos, E. E. 1982. Induced genetic changes in seed germplasm during storage. In A. A. Khan, ed., *The Physiology and Biochemistry of Seed Development, Dormancy, and Germination*, pp. 409–434. Amsterdam: Elsevier.

———. 1986. Precepts of successful seed storage. In M. B. McDonald, Jr., and C. J. Nelson, eds., *Physiology of Seed Deterioration*, pp. 1–25. Madison, Wis.: Crop Science Society of America Spec. Pub. 11.

———. 1989. Long-term seed storage. *Plant Breed. Rev.* 7:129–158.

———. 1992. Research program at the USDA National Seed Storage Laboratory. In A. A. Khan and J. Fu, eds., *Advances in the Science and Technology of Seeds*, pp. 153–163. Beijing: Science Press.

Roos, E. E. and D. A. Davidson. 1992. Record longevities of vegetable seeds in storage. *HortScience* 27:393–396.

Roos, E. E. and C. M. Rincker. 1982. Genetic stability in "Pennlate" orchardgrass seed following artificial aging. *Crop Sci.* 22:611–613.

Rupley, J. A., E. Gratton, and G. Careri. 1983. Water and globular proteins. *Trends in Biochem. Sci.* 8:18–22.

Sakai, A. 1986. Cryopreservation of germplasm of woody plants. In Y .P. S. Bajaj, ed., *Biotechnology in Agriculture and Forestry*. Vol 1: *Trees I*, pp. 113–129. Berlin: Springer-Verlag.

Sakai, A. and Y. Nishiyama. 1978. Cryopreservation of winter vegetative buds of hardy fruit trees in liquid nitrogen. *HortScience* 13:225–227.

Sedgley, M. 1981. Storage of avocado pollen. *Euphytica* 30:595–599.

Sivori, E., F. Nakayama, and E. Cigliano. 1968. Germination of Achira seed (*Canna* sp.) approximately 550 years old. *Nature* (London) 219:1269–1270.

Smith, R. D. 1992. Seed storage temperature and relative humidity (Correspondence). *Seed Sci. Res.* 2:113–116.

Spira, T. P. and L. K. Wagner. 1983. Viability of seeds up to 211 years old extracted from adobe brick buildings of California and Northern Mexico. *Am. J. Bot.* 70:303–307.

Stanwood, P. C. 1980. Tolerance of crop seeds to cooling and storage in liquid nitrogen (-196°C). *J. Seed Tech.* 5 (1): 26–31.

———. 1985. Cryopreservation of seed germplasm for genetic conservation. In K. K. Kartha, ed., *Cryopreservation of Plant Cells and Organs*, pp. 199–226. Boca Raton: CRC Press.

———. 1987. Survival of sesame seeds at the temperature (-196°C) of liquid nitrogen. *Crop Sci.* 27:327–331.

Stanwood, P. C. and L. N. Bass. 1981. Seed germplasm preservation using liquid nitrogen. *Seed Sci. Technol.* 9:423–437.

Stanwood, P. C. and E. E. Roos. 1979. Seed storage of several horticultural species in liquid nitrogen (-196°C). *HortScience* 14:628–630.

Steponkus, P. L., R. Langis, and S. Fujikawa. 1992. Cryopreservation of plant tissues by vitrification. In P. L. Steponkus, ed., *Advances in Low-Temperature Biology*, vol. 1, pp. 1–61. Greenwich: JAI Press.

Sur, K. and R. N. Basu. 1986. Control of age- and irradiation-induced seed deterioration in rice (*Oryza sativa* L.) by hydration-dehydration treatments. *Seed Res.* 14:197–205.

Tannoury, M., J. Ralambosoa, M. Kaminski, and J. Dereuddre. 1991. Cryoconservation by vitrification of alginate-coated carnation (*Dianthus caryophyllus* L) shoot tips of in vitro plantlets. *C. R. Acad. Sci. Paris* 313:633–638.

Tilden, R. L. and S. H. West. 1985. Reversal of the effects of aging in soybean seeds. *Plant Physiol.* 77:584–586.

Toole, E. H. and E. Brown. 1946. Final results of the Duvel buried seed experiment. *J. Agric. Res.* 72:201–210.

Toole, V. K. 1986. Ancient seeds: Seed longevity. *J. Seed Technol.* 10:1–10.

Towill, L. E. 1985. Low temperature and freeze-/vacuum-drying preservation of pollen. In K. K. Kartha, ed., *Cryopreservation of Plant Cells and Organs*, pp. 171–198. Boca Raton: CRC Press.

——. 1988. Genetic considerations for germplasm preservation of clonal materials. *HortScience* 23:91–97.

——. 1989. Biotechnology and germplasm preservation. *Plant Breed. Rev.* 7:159–182.

——. 1990a. Cryopreservation. In J. H. Dodds, ed., *In Vitro Methods for Conservation of Plant Genetic Resources*, pp. 41–70. London: Chapman and Hall.

——. 1990b. Cryopreservation of isolated mint shoot tips by vitrification. *Plant Cell Rep.* 9:178–180.

——. 1995. Vitrification techniques. In B. W. W. Grout, ed., *Plant Preservation in Vitro*, pp. 99–111. Berlin: Springer-Verlag.

Towill, L. E. and R. L. Jarret. 1992. Cryopreservation of sweet potato (*Ipomoea batatas* [L.] Lam.) shoot tips by vitrification. *Plant Cell Rep.* 11:175–178.

Towill, L. E. and E. E. Roos. 1989. Techniques for preservation of plant germplasm. In L. Knutson and A. K. Stoner, eds., *Biotic Diversity and Germplasm Preservation: Global Imperatives*, pp. 379–403. Dordrecht: Kluwer.

Tyler, N. and C. Stushnoff. 1988. Dehydration of dormant apple buds at different stages of cold acclimation to induce cryopreservability in different cultivars. *Can. J. Plant Sci.* 68:1169–1176.

Tyler, N., C. Stushnoff, and L. V. Gusta. 1988. Freezing of water in dormant vegetative apple buds in relation to cryopreservation. *Plant Physiol.* 87:201–205.

Uragami A, A. Sakai, M. Nagai, and T. Takahashi. 1989. Survival of cultured cells and somatic embryos of *Asparagus officinalis* cryopreserved by vitrification. *Plant Cell Rep.* 8:418–421.

Vertucci, C. W. 1989a. The effects of low water contents on physiological activities of seeds. *Physiol. Plant.* 77:172–176.

———. 1989b. Effects of cooling rate on seeds exposed to liquid nitrogen temperatures. *Plant Physiol.* 90:1478–1485.

———. 1989c. Relationship between thermal transitions and freezing injury in pea and soybean seeds. *Plant Physiol.* 90:1121–1128.

———. 1990. Seed germination. In *1991 Yearbook of Science and Technology*, pp. 374–377. New York: McGraw-Hill.

———. 1992. A calorimetric study of the changes in lipids during seed storage under dry conditions. *Plant Physiol.* 99:310–316.

———. 1993. Predicting the optimum storage conditions for seeds using thermodynamic principles. *J. Seed Technol.* 17 (2): 41–53.

Vertucci, C. W., P. Berjak, N. W. Pammenter, and J. Crane. 1991. Cryopreservation of embryonic axes of an homoeohydrous (recalcitrant) seed in relation to calorimetric properties of tissue water. *Cryo-Letters* 12:339–350.

Vertucci, C. W. and J. M. Farrant. 1995. Acquisition and loss of desiccation tolerance. In M. Negbi and J. Kigel, eds., *Seed Development and Germination*, pp. 237–271. New York: Marcel Dekker.

Vertucci, C. W. and E. E. Roos. 1990. Theoretical basis of protocols for seed storage. *Plant Physiol.* 94:1019–1023.

———. 1991. Seed moisture content, storage, viability, and vigour (correspondence). *Seed Sci. Res.* 1:277–279.

———. 1993a. Seed storage temperature and relative humidity (correspondence). *Seed Sci. Res.* 3:215–216.

———. 1993b. Theoretical basis of protocols for seed storage. Vol. 2: The influence of temperature on optimal moisture levels. *Seed Sci. Res.* 3:201–213.

Villiers, T. A. and D. J. Edgcumbe. 1975. On the cause of seed deterioration in dry storage. *Seed Sci. Technol.* 3:761–774.

Visser, T. 1955. Germination and storage of pollen. *Mededelinger van de Lanbouwhogeschool te Wageninger/Nederland* 55:1–68.

Vithanange, H. I. M. V. and D. Alexander. 1985. Synchronous flowering and pollen storage techniques as aids to artificial hybridization in pistachio (*Pistacia* spp.). *J. Hort. Sci.* 60:107–113.

Wanas, W. H., J. A. Callow, and L. A. Withers. 1986. Growth limitations for the conservation of pear genotypes. In L. A. Withers and P. G. Alderson, eds., *Plant Tissue Culture and Its Agricultural Applications*, pp. 285–290. London: Butterworths.

Webb, M. S., S. W. Hui, and P. L. Steponkus. 1993. Dehydration-induced lamellar to hexagonal II phase transitions in DOPE/DOPC mixtures. *Biochim. Biophys. Acta* 1145: 93–104.

Went, F. 1969. A long-term test of seed longevity. *Aliso* 7:1–12.

Went, F. and P. A. Munz. 1949. A long-term test of seed longevity. *Aliso* 2:63–75.

Wesley-Smith, J., C. W. Vertucci, P. Berjak, N. W. Pammenter, and J. Crane. 1992. Cryopreservation of desiccation-sensitive axes of *Camellia sinensis* in relation to dehydration, freezing rate, and the thermal properties of tissue water. *J. Plant Physiol.* 140:596–604.

Wester, H. V. 1973. Further evidence on age of ancient viable Lotus seeds from Pulantian deposit, Manchuria. *HortScience* 8:371–377.

White, G. A., H. L. Shands, and G. R. Lovell. 1989. History and operation of the National Plant Germplasm System. *Plant Breed. Rev.* 7:5–56.

White, J. 1909. The ferments and latent life of resting seeds. *Proc. Roy. Soc. London, Ser. B* 81:417–442.

Williams, R. J. and A. C. Leopold. 1989. The glassy state in corn embryos. *Plant Physiol.* 89: 977–981.

Wilson, D. O., Jr., and M. B. McDonald, Jr. 1986. The lipid peroxidation model of seed aging. *Seed Sci. Technol.* 14:269–300.

Withers, L. A. 1987. The low temperature preservation of plant cell, tissue, and organ cultures and seed for genetic conservation and improved agricultural practices. In B. W. W. Grout and G. J. Morris, eds., *The Effects of Low Temperatures on Biological Systems*, pp. 389–409. London: Arnold.

Withers, L. A. 1991. In vitro conservation. *Biol. J. Linnean Soc.* 43:31–42.

Yamaguchi, H., T. Naito, and A. Tatara. 1978. Decreased activity of DNA polymerase in seeds of barley during storage. *Japan. J. Genetics* 53:133–135.

Youngman, B. J. 1951. Germination of old seeds. *Kew Bull.* 6:423–426.

12

The Use of Herbarium Material for DNA Studies

DENNIS J. LOOCKERMAN AND ROBERT K. JANSEN

The use of herbarium specimens for molecular systematics studies is becoming more and more common. Microisolation of total DNA, using modifications of some common laboratory protocols, provides sufficient material for study with minimal specimen damage. Use of the polymerase chain reaction (PCR) and sequencing are the most common methods for producing data in these studies. We have obtained data for the genetic markers ndhF and ITS in the tribe Tageteae (Asteraceae) using herbarium specimens. We have also conducted surveys of both molecular systematics labs and herbaria in order to determine the extent of the use of herbarium collections and herbarium policies for sampling specimens. Molecular labs are increasing their usage of herbarium specimens and are examining DNA variation in a wide range of genetic markers in addition to the most common ones: rbcL, ndhF, and ITS. In general, herbaria are just beginning to implement policies regarding destructive sampling of specimens for molecular systematics studies, but they are amenable to discretionary usage.

During the past fifteen years there has been a surge in the use of DNA data for systematic and evolutionary studies of plants (Jorgensen and Cluster 1988; Palmer et al. 1988; Clegg 1993; Doyle 1993; Soltis, Soltis, and Doyle 1993). Because of the large quantities of leaf tissue required to extract sufficient quantities of DNA, most of these studies have examined DNA isolated from material collected in botanical gardens, the field, or greenhouses. Recent developments in molecular biology, however, most significantly the polymerase chain reaction (PCR), have enabled the use of very small amounts of DNA for molecular systematics and evolutionary studies. This new and powerful technique has revolutionized molecular systematics and enables biologists to examine DNA variation from very small amounts of plant tissue, even from museum collections (Rogers and Bendich 1985; Bruns, Fogel, and Taylor 1990; Blackwell and Chapman 1993; Sytsma et al. 1993; Taylor and Swann 1994), and from fossils (Golenberg et al. 1990; Soltis, Soltis, and Smiley 1992; Thomas and Paabo 1993).

In spite of recent technical advances in DNA isolation methods, the full potential of herbarium material has not been realized in molecular systematic comparisons. This is due, at least in part, to the absence of any published review on DNA methods for museum material or policies regarding accessibility of herbarium collections for use in molecular systematics labs. In this paper we will address five topics related to utilization of herbarium material for systematic studies involving DNA characters: (1) methods being used for DNA isolation; (2) examples of the use of herbarium material for phylogenetic comparisons; (3) extent of the use of herbarium specimens by molecular systematics labs and methods these labs are utilizing; (4) policies of herbaria for removing material from specimens; and (5) the potential of herbarium collections for future molecular investigations.

DNA Methods

DNA Characters for Systematic Comparisons

DNA characters for molecular systematics analysis fall into two general categories: structural changes and nucleotide substitutions. The chloroplast genome is more commonly used for systematic investigations involving structural changes. Common structural changes include inversions, gene losses or transfers, insertions/deletions, and changes in the size or structure of the inverted repeat region (Downie and Palmer 1992). Nucleotide substitutions can be examined indirectly by restriction site mapping (e.g., Jansen and Palmer 1988) or by randomly amplified polymorphic DNAs (RAPDs) or directly by DNA sequencing (Chase et al. 1993). RAPDs have limited use in phylogenetics, with their main application in population studies (e.g., Van Heusden and Bachmann 1992).

Total DNA Isolation

Methods of isolation of DNA from herbarium material generally consist of three steps: tissue grinding, organic extraction, and DNA precipitation. Since the quantity of material removed is generally small (less than 100 mg), the grinding step is often performed in a microcentrifuge tube. The most common method for DNA isolation uses the detergent hexadecyltrimethylammonium bromide (CTAB; Doyle and Doyle 1987) to break down membrane integrity and release the DNA. The organic compounds are then extracted one or more times with an organic solvent such as chloroform. Finally, the DNA is precipitated with alcohol (Doyle and Doyle 1987) or polyethylene glycol (PEG; Rowland and Nguyen 1993). A second method combines the extraction and precipitation steps. It uses a silica matrix to bind the DNA from the ground tissue. The matrix is then washed, and the DNA is eluted from the matrix (Cano and Poinar 1993).

The CTAB method is most commonly used (see table 12.1; Stein 1993) and is the method preferred in our lab. Here is the detailed protocol that we follow:

1. Adjust a water bath to 60°C and heat a sufficient quantity (ca. 1 ml/sample) of 2X CTAB with 1% sodium bisulfite and 1% polyvinylpyrrolidone (PVP, M.W. 40,000).
2. Weigh 30–50 mg of leaf tissue for each sample (highest DNA yields are obtained from the most recent collections and from the youngest, greenest leaves). Best results are obtained when tissue is rehydrated in double distilled (dd) H_2O while CTAB is heating. (Rehydration times are variable from a few minutes to overnight depending on the nature of the leaves. Thicker and waxy leaves require longer soaking.)

TABLE 12.1
Questions and Results of Molecular Systematics Lab Survey

1. Do you use DNA from herbarium specimens?
   Of 35 labs responding to the survey, 22 are currently using herbarium material. Most of those not using it expressed an interest in doing so in the near future.
2. What DNA isolation method do you use?
   Most labs (21) are using CTAB extractions (Doyle and Doyle 1987) modified for smaller quantities of tissue. A few labs modify the procedure by using 3X——6X CTAB and 2M NaCl. Other methods mentioned include Cano and Poinar 1993 (2 labs); Edwards, Johnstone and Thompson 1991 (1 lab); Gilmore, Weston, and Thomson 1993 (1 lab); and Rowland and Nguyen 1993 (1 lab). Our CTAB protocol can be found in the DNA Methods section.
3 What is the age of the specimens?
   From less than 1 to 99 years old, mostly less than or equal to 20 years old.
4. Which molecular approaches are you using?
   PCR and sequencing is most often used (19 labs) followed by PCR and restriction site analysis. Screening for inversions and gene loss, as well as for RAPDs, were also mentioned.
5. What percentage of DNA samples from your lab is derived from herbarium material?
   From 0 to 25% of samples per lab, but generally less than 10%.
6. Which taxonomic groups are being sampled?
   In Angiosperms, 25 orders from 7 subclasses of dicots and monocots were mentioned, as well as some mosses and ferns.
7. Which herbaria are providing specimens?
   The following herbaria were mentioned by the labs surveyed: ARIZ, BH, BISH, CAS, COLO, CT, CU, F, ILL, ISC, K, MO, MONT, NY, OSC, POM, RSA, TAMU, TEX, US, UTC, VT, WIS, WS. Most labs used the herbaria located in or near their own institutions.
8. Do you need to purify the DNA after isolation?
   Generally not. A couple of labs mentioned extra EtOH precipitations or gel purification. Glass-powder and resin-based purification kits were also mentioned.
9. Which molecular markers are being examined?
   Six chloroplast (*mat*K, *ndh*F, *psb*A, *rbc*L, *rpo*C1, and *rps*2), seven nuclear (ADH; IGS; ITS; 5.8s, 18s, and 26s rDNA; and glutamine synthase), and two mitochondrial markers (Cox I, Cox II) were mentioned. ITS, *rbc*L, and *ndh*F were the markers mentioned most often.

Labs responding: K. Bachmann, B. Baldwin, B. Bremer, J. Davis, C. DePamphilis, S. Downie, J. Doyle, B. Hahn, R. Jansen, J. Kadereit, S. Keeley, E. Kellogg, M. Lavin, D. Les, A. Liston, J. Manhart, C. Morden, D. Nickrent, D. Olmstead, J. Porter, R. Price, T. Ranker, L. Rieseberg, B. Schaal, J. Smith, D. Soltis and P. Soltis, D. Spooner, D. Stein, K. Sytsma, L. Urbatsch, R. Wallace, J. Wendel, P. Wolf, L. Zimmer.

3. After the CTAB has heated to 60°C and 0.2% betamercaptoethanol (BME) has been added, aliquot 0.2 ml of CTAB buffer into a "Kontes Duall" (Fisher Biotech) 1 ml tissue grinder, add rehydrated leaf tissue, and grind to a homogeneous paste.

4. Pour grindate into a 1.5 ml microcentrifuge tube, then add 0.6 ml CTAB buffer to the tissue grinder to rinse the remaining tissue into the 1.5 ml tube.

5. Place capped microcentrifuge tube into the 60°C water bath for fifteen minutes with occasional inverting.

6. Add 2/3 volume (0.55 ml) SEVAG (24:1 chloroform:octanol/isoamyl

alcohol) to each tube. Invert each tube several times to form an emulsion, occasionally releasing the gas between inversions.

7. Spin tubes in a microcentrifuge at full speed (ca. 12,000 rpm) for two minutes.

8. Carefully remove the aqueous (top) layer with a Pasteur pipette and place in a clean microcentrifuge tube.

9. Repeat steps 6–8 once more with the aqueous solution.

10. Add 2/3 volume (ca. 0.4 ml) ice-cold isopropanol to the aqueous solution and invert tubes several times.

11. Place tubes overnight in a -20°C freezer to allow DNA to precipitate.

12. After the DNA precipitates, adjust a water bath to 37°C. Spin tubes in a microcentrifuge for five minutes at full speed (ca. 12,000 rpm). If plug is floating or dispersed, spin for five more minutes.

13. Discard the supernatant and add 0.8 ml 76% EtOH/0.01 M NH_4OAc. Allow tubes to sit for ten minutes. Pour off supernatant.

14. Allow plug to air-dry, speedvac dry, or use a sterile Q-tip to remove any remaining isopropanol.

15. Resuspend pellet in 0.1 ml TE at 37°C for five to thirty minutes.

Polymerase Chain Reaction

Development of PCR has facilitated the use of herbarium specimens as sources of DNA. Without the technology for replicating specific segments of DNA, the minute quantity of total DNA obtainable from a herbarium specimen would be unusable. Recently, considerable literature has focused on optimizing PCR (e.g., Williams 1989; Bej, Mahbubani, and Atlas 1991; Cha and Thilly 1993). Below is a brief summary of the basic approach to performing PCR reactions.

The main components of a standard PCR reaction are template DNA, usually obtained from a total DNA isolation; two primers, one each for the 5´ and 3´ ends of the nucleotide sequence (generally oligonucleotides ca. 20 bp in length); a thermo-stable DNA polymerase; and the four deoxynucleotide triphosphates (dATP, dCTP, dGTP, and dTTP). These elements are combined with the proper buffers in a PCR tube and placed in a DNA thermocycler.

The DNA thermocycler controls the reaction conditions. The temperature and duration of each step is dependent on several factors: the length and complementarity of the primers, the length of the DNA seg-

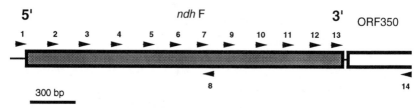

FIGURE 12.1   Chloroplast region containing *ndh*F (from Jansen 1992) showing PCR (1,8 and 7,14) and sequencing primers (1–14).

ment being amplified, and the G-C content of the primers and segments (see White et al. 1990). The following is a general algorithm we use to amplify subunit six of NADH dehydrogenase (*ndh*F, fig. 12.1) coded in the chloroplast genome and the Internal Transcribed Spacer (ITS, fig. 12.2) region of the nuclear ribosomal repeat. An initial denaturation step (ca. 95°C for 3 min) allows the DNA to become single-stranded; an annealing step (ca. 50°C for 1 min) enables the primers to bind to the target sequence; and a polymerization step (ca. 72°C for 1 min) allows the DNA polymerase to extend the DNA chains begun by the annealed primers. The subsequent 30 to 35 cycles reduce the denaturation time to one minute, but keep the other segments the same as the first cycle. At the end a polymerization step at 72°C for seven minutes is usually added to allow completion of all unfinished DNA strands (see Kim and Jansen 1994). The denaturation and annealing temperatures and times and the extension time can also vary greatly depending on the molecular marker and taxa involved.

PCR products can be used in several ways. DNA sequencing (e.g., Kim and Jansen 1994) and restriction enzyme analysis (e.g., Liston 1992) are the most common. Fragments are usually sequenced in one of three ways: cloning, in which a vector (such as M13) with the inserted DNA is cultured, and the replicated insert is removed and sequenced; single-stranded amplification, in which the PCR product is asymmetrically produced by limiting one of the primers, and the resulting single-stranded product sequenced (Gyllensten and Erlich 1988); and double-stranded sequencing, where the double-stranded PCR product is sequenced directly after heat denaturing the DNA (Winship 1989).

Restriction site analysis of PCR products generally uses longer DNA fragments (ca. 3–5 kb) than DNA sequencing (ca. 0.5–2 kb). It also employs restriction enzymes that cut more frequently (4-base pair recognition sequence) than those generally used for genomic DNA restriction

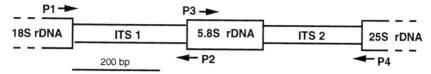

FIGURE 12.2.   Nuclear ribosomal Intercistronic Transcribed Spacer (ITS) region showing PCR (P1, P4) and sequencing (P1, P2, P3, and P4) primers (from Kim and Jansen 1994).

site analyses (6-base pair recognition sequence). Since the reaction conditions for PCR become more critical as the length of the target sequence increases, and since the number of possible characters available for a restriction analysis is less than the number available for DNA sequencing, the main use of PCR products is DNA sequencing. This is especially true when using herbarium material because the DNA is likely to be degraded making amplification of longer sequences more difficult.

Suggestions for PCR of Microextracted Herbarium Material

We generally resuspend the microextracted DNA in TE buffer. Since EDTA is a chelating agent we have optimized the $MgCl_2$ concentration experimentally in the following manner: multiple samples of the DNA are run with the $MgCl_2$ concentration being varied from 1 to 4 mM in 0.5 mM increments, the sample that has the best fragment amplification with the least background amplification is the one used in future PCR runs. We first perform PCRs of new material using 25 µl reactions. Using 1 µl of total DNA for each reaction and three dilutions of the microextracted DNA of 1X, 0.1X, and 0.01X. This approach results in approximately a 50% success rate for one of these dilutions working on the first attempt.

If none of the initial dilutions works, samples of the original DNA are run on a minigel to inspect the quality of the extracted DNA. If the extracted DNA seems weak or undetectable, the amount of DNA in the PCR reaction (2 to 5 µl and sometimes up to 10 µl) can be increased. This usually enables the amplification of DNA from another 25% of the taxa.

If the minigel shows sufficient DNA in the original extraction (and especially if the DNAs are colored) we use a glass-powder-based cleaning kit such as "Geneclean II" (BIO 101) to purify the DNA from compounds that may be interfering with the PCR reaction. Alternatively, the DNA can be purified using ethanol precipitation (Maniatis, Fritsch, and Sambrook 1989). This normally enables the remaining 25% of the samples to be amplified. In several cases the previous series of steps has not proven successful in our lab; in these instances we simply extracted DNA from a different specimen and performed the above procedures. We have extracted DNA from specimens ten to twenty years old routinely and have had success with specimens up to ninety-nine years old (D. Loockerman, unpublished).

Examples of the Use of Herbarium Material for Phylogenetic Studies

Our lab is currently conducting a number of molecular systematics studies that employ DNA isolated from herbarium specimens. Some of these use DNA primarily extracted from fresh leaf material with herbarium specimens used to supplement DNA from species that are not otherwise accessible (e.g., Cosner, Jansen, and Lammers 1994). Two large studies are being based primarily or exclusively on herbarium specimens.

PCR of DNA fragments is being carried out as explained in previous sections. We are using direct double-stranded DNA sequencing by denaturing the template in 95°C water for three minutes and then snap-chilling in an ice bath for five minutes. The remainder of the protocol follows standard Sequenase (U.S. Biochemicals) procedures (see also Winship 1989; Thomas and Paabo 1993).

Our most extensive project to date that is relying primarily on herbarium specimens is a phylogenetic and molecular evolutionary investigation of the tribe Tageteae (Asteraceae). The Tageteae comprise 240 species (Strother 1977, 1986) divided among 16 to 23 genera. The tribe belongs in subfamily Asteroideae near tribe Heliantheae (Kim and Jansen, 1996).

The goals of this study are to determine the systematic relationships in the tribe and to analyze rates and patterns of DNA sequence evolution. Two molecular markers are being examined to accomplish these goals: ndhF and ITS. The majority of the DNA used in this study is being obtained from herbarium specimens.

The 3′ portion of the ndhF gene (between primers 7 and 14, fig. 12.1) has been sequenced for 34 of the proposed 51 taxa, and 32 of these used DNA from herbarium specimens. Of the proposed 51 taxa, 2 have been

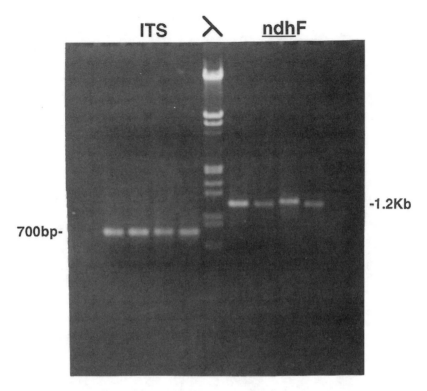

Figure 12.3. Sample minigel showing ITS (left four lanes) and *ndh*F (right four lanes) PCR products using DNA isolated from herbarium specimens. PCR products were amplified using primers P1 and P4 (fig. 12.2) for ITS and primers 7 and 14 (fig. 12.1) for *ndh*F using protocols explained previously in text. Products were run out on a 1.5% agarose gel and stained with ethidium bromide.

sequenced for ITS (fig. 12.2) and 2 of those from herbarium specimens. The data and results for this study will be published at a later date as part of Dennis J. Loockerman's Ph.D. dissertation.

The second study we are conducting is an intergeneric study of the tribe Mutisieae (Asteraceae). This study will examine ca. 100 taxa, all from herbarium specimens. This group is proving to be much more difficult to amplify DNA from herbarium specimens. Thus our success rate is only about 25%.

Representative samples of PCR products (fig. 12.3) and a sequencing gel (fig. 12.4) show the quality and quantity of DNA fragments that we are obtaining in these studies.

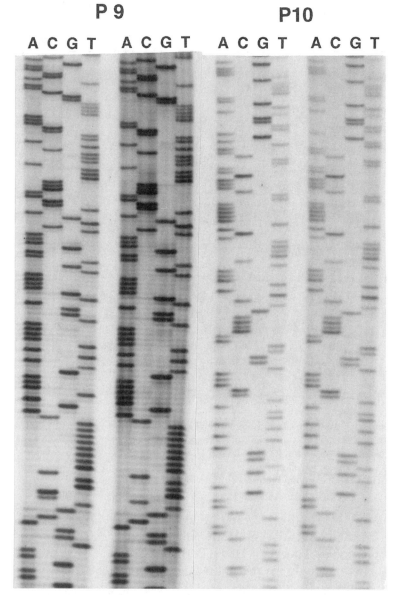

FIGURE 12.4 Sample sequencing gel of the *ndh*F region using DNA isolated from herbarium specimens. P9 and P10 are two of the sequencing primers shown in figure 12.1. Double-stranded PCR products were sequenced as explained in the text. Sequencing products were heat-denatured and then run at 1,400 volts on a 6% polyacrylamide denaturing gel.

Use of Herbarium Material by Molecular Systematics Labs

We conducted a survey of forty-three molecular systematics labs to determine the extent to which herbarium material is currently being used for DNA studies (table 12.1). Most labs are using at least some herbarium specimens, although generally it is less than 10% of the total samples analyzed. CTAB extraction (Doyle and Doyle 1987) is the preferred method of DNA isolation, and PCR with sequencing is the most widely used systematic approach. Several chloroplast, several nuclear, and a few mitochondrial markers are being examined, the most frequently used being the nuclear ribosomal ITS region and the chloroplast markers *rbcL* and *ndhF*. The taxa being examined represent seven subclasses of dicots and monocots in the Angiosperms and some mosses and ferns.

Herbarium Policies for DNA Studies

We also conducted a survey of fifty-five herbaria to ascertain the policies and amenability of these institutions toward extracting DNA from herbarium specimens (table 12.2). In general, most herbaria allow removal of material with prior consent, assuming that the specimen is not noticeably degraded. The demand for specimens for DNA studies is currently very low but is increasing. Material is generally not allowed to be removed from type specimens, although some institutions will allow its removal in special cases. Acknowledgment of the herbarium is most often requested by citation of the voucher specimen, including the herbarium acronym.

Conclusions and Future Prospects

It is clear from our survey that the use of herbarium specimens for DNA studies is rising, and may soon become a significant proportion of the DNA resources for some labs (Blackwell and Chapman 1993; Sytsma et al. 1993). Herbarium collections are important for molecular studies for several reasons: (1) they provide automatic vouchers; (2) the collections can be used to complete missing data that were not easily obtainable (e.g., Cosner, Jansen, and Lammers 1994); (3) they can be used as the main source of data for widespread taxonomic groups for which field

TABLE 12.2
Survey of Herbarium Policies for Using Specimens in Molecular Investigations

1. Do you allow small amounts of material (e.g., 25 mg of leaves) to be removed from herbarium sheets for DNA studies?
   Generally yes (25/28) and sometimes only under certain circumstances (2/28) or no (1/28). It should be noted that material removed should be taken from excess-material envelopes (when available) and that no degradation of the specimen, for future morphological studies, should occur. Also, complete voucher labels should be filled out.
2. Do you have a policy with regard to removing material from specimens?
   Yes (25/28). We do now (3/28).
   A. Do you only allow removal of material by obtaining permission on a sheet-by-sheet basis, or would you give permission to remove material from any specimen on loan?
   On a sheet-by-sheet basis (14/28). On any specimen that has sufficient material (7/28). It depends on the person making the request (6/28). No removal of material (1/28).
   B. Do you fill requests for material for molecular studies by providing leaves and accompanying data (but not the specimens)?
   Yes (11/27). No (10/27). We have never had such a request, but we would comply (6/27).
3. What is the demand for dried material for DNA studies at your institution?
   Very low or none (25/28). Low but growing (3/28).
4. What percentage of loans request permission to remove leaf material for DNA studies?
   0% (18/28); 1—2% (6/28); 5% (3/28); 5—10% (1/28). Note: Most requests at present come from within the same institution where the herbarium is located.
5. Do you allow leaf material from types to be removed?
   No (16/28). Only on special request, assuming the specimen has sufficient material to prevent degradation (12/28).
6. How would you like your facility acknowledged when herbarium material is used for DNA studies?
   Cite the voucher specimen and include the acronym of the herbarium (26/28). Acknowledge the curator of the loaning institution (4/28). *Note*: A reprint of the paper was requested by several herbaria.
7. Plant specimens are sometimes preserved with alcohol or formalin or both when drying facilities are not available in the field. Since these chemicals might affect molecular analysis, do you make sure that information about the drying method is noted on the label?
   No (15/26). Make a best effort to do so (6/26). Have not, but will try to in the future (5/26). Note: It was mentioned by numerous herbaria that since much material is obtained by exchange, the above information is not available.

Special Comments

The following are special comments that herbaria requested be published with the survey.
1. Perhaps herbaria should request that the GENbank accession number be put on the voucher labels for any DNA studies (as well as other pertinent information).
2. Fumigation with methyl bromide and ultra-freezing are common practices. What effects do these processes have on DNA recovery? Ultra-freezing has not been found to be harmful to DNA recovery and is a common step in many DNA isolation procedures.
3. If DNA sampling becomes common practice, care needs to be taken so that rare specimens are not oversampled and that subsequent workers refer to published sequences when available.

Herbaria responding: ASU, B, BISH, BM, C, CAS, COLO, DAO, E, F, FSU, G, GA, GH, LINN, MA, MEXU, MICH, NY, OSC, P, PH, RSA SEL, TAMU, TEX, UC, WIS.

work might be impracticable (the Tageteae study mentioned previously); (4) the extensive museum collections worldwide enable the examination of a high proportion of species without field collections; and (5) it is possible to include rare and endangered, or even extinct, species in molecular systematics studies.

Our questionnaire has prompted several herbaria to reevaluate their policies with respect to destructive sampling. In the future, herbaria may request that some or all of the following guidelines be followed: submission of a proposal, including the DNA isolation protocol, with the material request; removal of material from a specific specimen for DNA studies may be limited to one-time only; GenBank/EMBL accession numbers may need to be included on voucher labels; samples of DNA, for storage by the lending institution, may be required; and copies (in writing) of all results and publications may be requested.

Herbaria are amenable to having their collections used for DNA studies, but this puts substantial responsibility on molecular systematists who will use this resource. Proper voucher labels should be made and GenBank accession numbers included so that unnecessary duplication of effort and oversampling do not occur. Herbaria should be acknowledged in any publications; this is especially important today since many institutions may be contemplating the need to maintain vast herbarium collections. In addition, continued collecting of new specimens should still be encouraged. Despite the availability of millions of specimens in the world's herbaria, many groups are not well sampled. Additional collections must sample underrepresented taxa, provide multiple collections for population studies, and provide recent collections that are more likely to yield high-quality DNA.

Acknowledgments

We thank the herbarium curators and heads of molecular systematics labs for generously providing information for our surveys (see tables 12.1 and 12.2), the University of Texas Herbarium (TEX), Carol Todzia for her help in designing the herbarium survey, and John Averett, Todd Barkman, Alice Hempel, Ki-Joong Kim, and Carol Todzia for their critical reading of the manuscript. This research was supported by NSF grants to R.K.J. (DEB-9318279) and to D.J.L. (DEB-9411536).

Literature Cited

Bej, A. K., M. H. Mahbubani, and R. M. Atlas. 1991. Amplification of nucleic acids by polymerase chain reaction (PCR) and other methods and their applications. *Crit. Rev. Biochem. Mol. Biol.* 26:301–334.

Blackwell, M. and R. L. Chapman. 1993. Collection and storage of fungal and algal samples. In E. A. Zimmer, T. J. White, R. L. Cann, and A. C. Wilson, eds., *Methods in Enzymology,* vol. 224, pp. 65–77. San Diego: Academic Press.

Bruns, T. D., R. Fogel, and J. W. Taylor. 1990. Amplification and sequencing of DNA from fungal herbarium specimens. *Mycologia* 82:175–184.

Cano, R. J. and H. N. Poinar. 1993. Rapid isolation of DNA from fossil and museum specimens suitable for PCR. *Biotechniques* 15:432–435.

Cha, R. S. and W. G. Thilly. 1993. Specificity, efficiency, and fidelity of PCR. *PCR Methods and Applications,* Manual Supplement, pp. S18–S29. Cold Spring Harbor: Cold Spring Harbor Laboratory.

Chase, M., D. Soltis, R. Olmstead, D. Morgan, D. Les, B. Mishler, M. Duvall, R. Price, H. Hills, Y.-l. Qui, K. Kron, J. Rettig, E. Conti, J. Palmer, J. Manhart, K. Sytsma, H. Michaels, J. Kress, K. Karol, D. Clark, M. Hedren, B. Gaut, R. Jansen, K.-J. Kim, C. Wimpee, J. Smith, G. Furnier, S. Straus, Q.-y. Xiang, G. Plunkett, P. Soltis, S. Swensen, S. Williams, P. Gadek, C. Quinn, L. Equiarte, E. Golenberg, G. Learn, Jr., S. Graham, S. Barrett, S. Dayanandan and V. Albert. 1993. Phylogenetics of seed plants: An analysis of nucleotide sequences from the plastid gene *rbc*L. *Ann. Missouri Bot. Gard.* 80: 528–580.

Clegg, M. T. 1993. Chloroplast gene sequences and the study of plant evolution. *Proc. Natl. Acad. Sci. USA* 90:363–367.

Cosner, M. E., R. K. Jansen, and T. G. Lammers. 1994. Phylogenetic relationships in the *Campanulales* based on *rbc*L sequences. *Pl. Syst. Evol.* 190:79–95.

Downie, S. R. and J. D. Palmer. 1992. Use of chloroplast DNA rearrangements in reconstructing plant phylogeny. In P. Soltis, D. Soltis, and J. Doyle, eds., *Molecular Systematics of Plants,* pp. 14–35. New York: Chapman and Hall.

Doyle, J. J. and J. L. Doyle. 1987. A rapid DNA isolation procedure for small quantities of fresh leaf tissue. *Phytochem. Bull.* 19:11–15.

——. 1993. DNA, phylogeny, and the flowering of plant systematics. *BioScience* 43:380–390.

Edwards, K., C. Johnstone, and C. Thompson. 1991. A simple and rapid method for the preparation of plant genomic DNA for PCR analysis. *Nucleic Acids Res.* 19:1349.

Gilmore, S., P. H. Weston, and J. A. Thomson. 1993. A simple, rapid, inexpensive and widely applicable technique for purifying plant DNA. *Austral. Syst. Bot.* 6:139–148.

Golenberg, E. M., D. E. Giannasi, M. T. Clegg, C. J. Smiley, M. Durbin, D. Henderson, and G. Zurawski. 1990. Chloroplast DNA sequence from a Miocene *Magnolia* species. *Nature* 344:656–658.

Gyllensten, U. B. and H. A. Erlich. 1988. Generation of single-stranded DNA by the polymerase chain reaction and its application to direct sequencing of the HLA-DQA locus. *Proc. Natl. Acad. Sci. USA* 85:7652–7656.

Jansen, R. K. and J. D. Palmer. 1988. Phylogenetic implications of chloroplast

DNA restriction site variation in the Mutisieae (Asteraceae). *Amer. J. Bot.* 75:751–764.

———. 1992. Current research. *Pl. Mol. Evol. Newsl.* 2:13–14.

Jorgensen, R. A. and P. D. Cluster. 1988. Modes and tempos in the evolution of nuclear ribosomal DNA: New characters for evolutionary studies and new markers for genetic and population studies. *Ann. Missouri Bot. Gard.* 75:1238–1247.

Kim, K.-J. and R. K. Jansen. 1994. Comparisons of phylogenetic hypotheses among different data sets in dwarf dandelions (*Krigia, Asteraceae*): Additional information from internal transcribed spacer sequences of nuclear ribosomal DNA. *Pl. Syst. Evol.* 190:157–185.

———. 1996. *ndh*F sequence evolution and the major clades in the sunflower family. *Proc. Natl. Acad. Sci. USA.* 92:10379–10383.

Liston, A. 1992. Variation in the chloroplast genes *rpo*C1 and *rpo*C2 of the genus *Astragalus* (Fabaceae): Evidence from restriction site mapping of a PCR-amplified fragment. *Amer. J. Bot.* 79:953–961.

Maniatis, T., E. F. Fritsch, and J. Sambrook. 1989. *Molecular Cloning: A Laboratory Manual*, 2d ed. Cold Spring Harbor: Cold Spring Harbor Laboratory.

Palmer, J. D., R. K. Jansen, H. J. Michaels, M. W. Chase, and J. R. Manhart. 1988. Chloroplast DNA variation and plant phylogeny. *Ann. Missouri Bot. Gard.* 75:1180–1206.

Rogers, S. O. and A. J. Bendich. 1985. Extraction of DNA from milligram amounts of fresh, herbarium, and mummified plant tissues. *Plant Mol. Biol.* 5:69–76.

Rowland, L. J. and B. Nguyen. 1993. Use of polyethylene glycol for purification of DNA from leaf tissue of woody plants. *Biotechniques* 14:735–736.

Soltis, P. S., D. E. Soltis, and J. Doyle, eds. 1993. *Molecular Systematics of Plants.* New York: Chapman and Hall.

Soltis, P. S., D. E. Soltis, and C. J. Smiley. 1992. An *rbc*L sequence from a Miocene *Taxodium* (bald cypress). *Proc. Natl. Acad. Sci. USA* 89:449–451.

Stein, D. B. 1993. Isolation and comparison of nucleic acids from land plants: Nuclear and organellar genes. In E. A. Zimmer, T. J. White, R. L. Cann, and A. C. Wilson, eds., *Methods in Enzymology*, vol. 224, pp. 153–167. San Diego: Academic Press.

Strother, J. 1977. Tageteae–systematic review. In V. H. Heywood, J. D. Harborne, and B. L. Turner, eds., *The Biology and Chemistry of the Compositae*, pp. 769–783. London: Academic Press.

———. 1986. Renovation of *Dyssodia* (Compositae: Tageteae). *Sida* 11:371–378.

Sytsma, K. J., T. J. Givnish, J. F. Smith, and W. J. Hahn. 1993. Collection and storage of land plant samples for macromolecular comparisons. In E. A. Zimmer, T. J. White, R. L. Cann, and A. C. Wilson, eds., *Methods in Enzymology*, vol. 224, pp. 23–37. San Diego: Academic Press.

Taylor, J. W. and E. C. Swann. 1994. DNA from herbarium specimens. In B. Herrmann and S. Hummel, eds., *Ancient DNA: Recovery and analysis of genetic material from paleontological, archaeological, museum, medical, and forensic speci-mens*, pp. 166–181. New York: Springer-Verlag.

Thomas, W. K. and S. Paabo. 1993. DNA sequences from old tissue remains. In E. A. Zimmer, T. J. White, R. L. Cann, and A. C. Wilson, eds., *Methods in Enzymology*, vol. 224, pp. 407–419. San Diego: Academic Press.

Van Heusden, A. W. and K. Bachmann. 1992. Genetic differentiation of *Microseris pygmaea* (Asteraceae, Lactuceae) studied with DNA amplification from arbitrary primers (RAPDs). *Acta Bot. Neerl.* 41:385–395.

White, T. J., T. Bruns, S. Lee, and J. Taylor. 1990. Amplification and direct sequencing of fungal ribosomal RNA genes for phylogenetics. In M. Innis, D. Gelfand, J. Sninsky, and T. White, eds., *PCR Protocols*, pp. 315–322. San Diego: Academic Press.

Williams, J. F. 1989. Optimization strategies for the polymerase chain reaction. *BioTechniques* 7:762–768.

Winship, P. 1989. An improved method for directly sequencing PCR amplified material using dimethyl sulphoxide. *Nucleic Acids Res.* 17:1266.

13

DNA Data Banking and Its Role in Documenting Phytodiversity

ROBERT P. ADAMS

The development of the polymerase chain reaction (PCR) now makes it possible to amplify DNA from DNA preserved under various conditions. With the anticipated loss of numbers of tropical plant species, DNA data banking has been proposed to document the diversity found in nature. An overview of the formation and function of DNA Bank-Net is presented. DNA Bank-Net is an informal association of institutions that have begun to accession DNA or DNA-rich materials for subsequent disbursement of genes or sequences. DNA Bank-Net should complement activities already being performed by different institutions, specifically those working in the area of germplasm collection and conservation.

Collections of plant specimens have been utilized to formulate an understanding of morphological variation among taxa. Indeed, without the great herbaria of the world, our knowledge of plant evolution would be highly fragmented. As we have moved to use chemical data for systematic and evolutionary studies, methods of preserving plant materials for future (chemical) work have been largely ignored. We are usually content simply to file a voucher specimen to document our chemical studies. With the present level of support for plant collections, it is unlikely that much of the world's plant species can be preserved by freezing so that scientists might have access to the study of secondary compounds, enzymes, or DNA/RNA in the coming centuries.

Since the first plant-to-plant gene transfer in 1983 (Murai et al. 1983), genes have been transferred to plants from viruses (Nelson et al. 1988), bacteria (Barton, Whiteley, and Yang 1987; Della-Cioppa et al. 1987; Fischhoff et al. 1987), and even from mammals to plants (Lefebvre, Miki, and Laiberte 1987; Maiti, Hunt, and Wagner 1988). Genetic transfers are being performed in order to attain insect, bacterial, viral, and fungal resistance, a more nutritionally balanced protein, more efficient photosynthesis, nitrogen fixation, and salt and heavy metal tolerance, to name a few. These kinds of gene transfers from one unrelated organism to another indicate that we must now view the world's genetic resources (genes, DNA) from a horizontal perspective in which gene transfers will cut across species, genera, and family boundaries.

For example, a strain of cowpea, *Vigna unguiculata* (L.) Walp., discovered in a market in Ilorin, Nigeria, contains a protein that inhibits trypsin digestion by insects (Redden, Singh, and Luckefahr 1984). This gene has been moved to tobacco (*Nicotiana*) where the trypsin-inhibiting gene is expressed and offers tobacco the same resistance against insects as in cowpea (Newmark 1987). It is interesting to note that although a very active form of the gene has been found in a Nigerian cowpea, scarcely 100 of the world's 13,000 legume species have been examined for this gene. Yet the tropical legumes, one of the most promising groups for the evolution of natural insecticides, will certainly be subject to considerable germplasm loss in the next decade.

The number of novel insecticides, biocides, medicines, and so on, that could exist in nature is innumerable. Yet the principal areas of diversity among plants, the lowland tropical forests, will have been cut or severely damaged within the next twenty years (Raven 1988). The Amazon River system, for example, contains eight times as many species as the Mississippi River system (Shulman 1986). Raven (1988) estimated that as many as 1.2 million species would become extinct in the next twenty years. The loss of plant species will mean a loss of potential plant-derived pharmaceuticals, now estimated at $2 billion per year in the United States alone (U.S. Congress 1987).

The loss of native plant diversity also means a loss of genetic diversity present in and available to our current and potential crop species. Cultivated crops are extremely inbred for factors such as yield, uniform flowering and height, and cosmetic features of products. This narrow genetic base has resulted in several disastrous crop failures. Ireland's potato (*Solanum tuberosum* L.) famine of 1846, which resulted in the emigration of a quarter of the human population, was because their potatoes had no resistance to the late blight fungus [*Phytophthora infestans* (Mont.) DeBary] (Plucknett et al. 1987). This can be traced to the lack of genetic diversity in Irish potatoes, which had been multiplied using clonal materials from just two separate South American introductions to Spain in 1570 and to England in 1590 (Hawkes 1979).

A more recent example is the southern corn leaf blight (fungus, *Helminthosporium maydis* Nisik. and Miy.) in 1970 in the United States. Because almost all the corn (*Zea mays* L.) in the United States was of hybrid origin and contained the Texas cytoplasmic male sterile line, our fields of corn presented an unlimited, extremely narrow gene base habitat for the fungus. By late in the summer of 1970 plant breeders were scouring corn germplasm collections in Argentina, Hungary, Yugoslavia, and the United States for resistant sources (Plucknett et al. 1987). Nurseries and seed fields were used in Hawaii, Florida, the Caribbean, and Central and South America to incorporate the resistance into hybrid corn in time for planting in the spring of 1971 (Ullstrup 1972). Without these genetic resources this technological feat would not have been possible.

The National Cancer Institute (NCI) is now spending $8 million over the next five years for a massive plant-collecting effort in the tropics to find anticancer and anti-AIDS virus compounds (Booth 1987; see also chapter 5). The plant collectors will gather leaves or bark or both and air-dry the material for shipment to Maryland where it will be extracted and assayed against a hundred cancer cell lines and the AIDS virus. Yet *no* genetic resources will be collected! When a promising compound is found, the plants will have to be recollected. For extensive testing (as well as for commercial utilization), plantations will have to be established in the tropics to provide material.

Formation of DNA Bank-Net

Concurrent with the advancements in gene cloning and transfer has

DNA Bank–Net

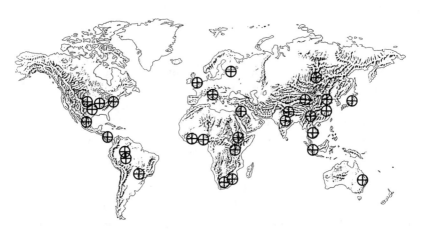

Figure 13.1   Map of individuals/institutions currently interested in DNA Bank-Net.

been the development of technology for the removal and analyses of DNA. DNAs from the nucleus, mitochondrion, and chloroplast are now routinely extracted and immobilized onto nitrocellulose sheets where the DNA can be probed with numerous cloned genes. Recent advances in the technology for the extraction and immobilization of DNA, coupled with the prospect of the loss of significant plant genetic resources throughout the world, has led to the establishment of DNA Bank-Net, an international network of DNA repositories for the storage of genomic DNA on every continent.

A group of eighteen scientists held the organizational meeting at the Royal Botanic Gardens, Kew, London, in April 1991 to share country and institutional experiences using in vitro biotechnology and particularly cryostorage of DNA and DNA-rich materials (Adams and Adams 1991). Relatively few scientists were interested in a "genetic insurance policy" when the idea of banking genomic DNA from plants was first proposed (Adams 1988, 1990). Currently, however, forty institutions (representing twenty-five nations and every continent, see table 13.1) have expressed interest in DNA Bank-Net (fig. 13.1).

The conserved DNA will have numerous uses: molecular phylogenetics and systematics of extant and extinct taxa; production of previously characterized secondary compounds in transgenic cell cultures;

TABLE 13.1
Individuals/Institutions That Have Expressed an Interest in DNA Bank-Net

Dr. Daniel K. Abbiw, Botany Department, University of Ghana
    Box 55, Legon, Ghana, West Africa
Dr. Robert P. Adams. DNA Bank-Net Director, Plant Biotechnology Center
    P.O. Box 97372, Baylor University, Waco, TX 76798
Dr. Vishwanath P. Agrawal, Research Lab for Agric. Biotechnology and
    Biochemistry, P.O. Box 2128, Kathmandu, Nepal
Drs. Lucia Atehortua/Ricardo Callejas, HUA, Department of Biology
    University of Antioquia, Medellin, Colombia
Dr. Luiz Antonio Barreto de Castro, CENARGEN/EMBRAPA
    Parque Rural, CP 102372, W 70770, Brasilia DF, Brazil
Dr. Michael Bennett, Director, Jodrell Labs, Royal Botanic Gardens
    Kew, Richmond, Surrey, TW9 3AB, England
Dr. Brian M. Boom, Vice President for Botanical Science
    The New York Botanical Garden, Bronx, NY 10458
Dr. Anthony H. D. Brown, Division of Plant Industry, CSIRO
    GPO Box 1600, Canberra, ACT 2601, Australia
Dr. Ram Chaudhary, Research Lab. for Agricultural, Biotechnology, and
    Biochemistry, P.O. Box 2128, Kathmandu, Nepal
Prof. Cheng Xiongqying, Institute of Nuclear Agricultural Sciences
    Zhejiang Agricultural University, Hangzhou, 310029 China PRC
Prof. J. Eloff, Director, National Botanical Institute
    Private Mail Bag X101, Pretoria, 0001 South Africa
Dr. Z. O. Gbile, Director, Forestry Research Institute of Nigeria
    Private Mail Bag 5054, Ibaden, Nigeria
Dr. Chaia C. Heyn, Department of Botany
    The Hebrew University, Jerusalem, 91904, Israel
Dr. Toby Hodgkin, Research Officer, IBPGR, c/o FAO of the UN
    Via delle Sette Chiese 142, 00145 Rome, Italy
Prof. Hu Zhong, Kunming Institute of Botany,
    The Academy of Sciences of China, Heilongtan, Kunming, Yunnan,
    China PRC
Dr. Kunio Iwatsuki, Botanical Gardens, University of Tokyo
    3-7-1 Hakusan, Bunkyo-ku, Tokyo 112, Japan
Prof. Mupinganayi Kadiakuida, Director General, CARI
    B.P. 16513, Kinshasa, Republic of Zaire
Dr. S. L. Kapoor, Cytogenetics Lab, National Botanical Research Institute
    Rana Pratap Marg., Lucknow 226 001, India
Prof. Lin Zhong-ping, Division of Plant Molecular Biology
    Institute of Botany, Academia Sinica, Beijing 100044, China PRC

Dr. Ma Cheng, Chief Engineer, Chinese Academy of Sciences
(Academia Sinica), 52 San Li He Road, Beijing, China PRC
Dr. Lydia Makhubu, University of Swaziland
Kwaluseni Campus P/Bag, Kwaluseni, Swaziland
Dr. John S. Mattick, Centre for Molecular Biology and Biotechnology
University of Queensland, St. Lucia, Queensland, QLD 4072, Australia
Dr. James Miller, Missouri Botanical Garden
2345 Tower Grove Ave., St. Louis, MO 63110
Dr. Titus K. Mukiama, Department of Botany, University of Nairobi
Chiromo, P.O. Box 30197, Nairobi, Kenya
Dr. Alfred Apau Oteng-Yeboah, Dept. of Botany, University of Ghana
P.O. Box 55, Legon, Ghana
Dr. Bart Panis, Laboratory of Tropical Crop Husbandry, Catholic University
of Leuven, Kardinaal Mercierlaan 92, B-3001 Heverlee, Belgium
Dr. Ghillean T. Prance, Director, Royal Botanic Gardens
Kew, Richmond, Surrey, TW9 3AB, England
Dr. Steve Price, University- Industry Research Program, 1215 WARF Bldg.
610 Walnut St., University of Wisconsin, Madison, WI 53705
Dr. Wickneswari Ratnam, Forestry Research Institute, Karung Berkunci 201
JLN FRI Kepong, 52109, Kuala Lumpur, Malayasia
Dr. Loren Rieseberg, Indiana University
Biology Dept., Bloomington, IL 47405
Dr. W. Roca, Head, Bio-Tech Unit
CIAT AA6713, Cali, Colombia
Dr. Phillip Stanwood, National Seed Storage Lab, USDA
Colorado State University, Ft. Collins, CO 80523
Dr. Dennis Stevenson, The New York Botanical Garden
Bronx, NY 10458
Dr. Peter Strelchenko, N. I. Vavilov Institute of Plant Industry
42 Herzen Street, 190000, St. Petersburg, Russia
Dr. Panie Temiesak, Head, Seed Technology, Central Lab and Greenhouse
Complex, Kasetsart University, Nakorn Pathom 73140, Thailand
Dr. Victor M. Villalobos, FAO/AGP, Via delle Terme di Caracalla
0010 Rome, Italy
Prof. Luz Ma. Villarreal de Puga, Instituto de Botanica University de
Guadalajara, Apartado 139, Las Agujas Nextipac, Zapopan, Jalisco, Mexico
Dr. Melaku Worede, Plant Genetic Resources Center
P.O. Box 30726, Addis Ababa, Ethiopia
Prof. Zheng Sijun, Department of Agronomy, Zhejiang Agricultural
University, Hangzhou, 310029, Zhejiang, China PRC
Prof. Zhu Ge-lin, Institute of Botany, Northwest Normal University
Lanzhou, Gansu, China PRC 730070

production of transgenic plants using genes from gene families; in vitro expression and study of enzyme structure and function; and genomic probes for research laboratories.

Structure and Operation of DNA Bank-Net

At the organizational meeting of DNA Bank-Net, a task force was convened to define the functions of working (DNA-dispensing) and reserve (base) nodes in the DNA bank network. The group recommended the following functions (Adams and Adams 1991):

Working (DNA-dispensing) nodes

a. Collection of plant material by taxonomists. This may be the primary function of a particular node or be in association with other organizations such as universities, botanic gardens, and so on.
b. DNA extraction by molecular biologists or trained staff.
c. Long-term preservation of DNA-rich materials and/or extracted DNA in liquid nitrogen.
d. DNA analysis/gene replication by molecular biologists or trained staff.
e. Distribution of DNA (genes, gene segments, oligonucleotides, etc.).

Reserve (base) nodes

a. Long-term DNA preservation in liquid nitrogen and monitoring of potential DNA degradation.
b. Act as genetic reserve buffer for working nodes.
c. Replenishment of DNA if a working node experiences the catastrophic loss of storage parameters and DNA.

Figure 13.2 depicts the relationship between working and reserve nodes. Note the projected flow of plant materials and DNA through the working (DNA-dispensing) node. It is likely that some of the working nodes would be actively acquiring or dispensing DNA or both from some geographic area (e.g., Africa) and yet maintain separate cryovats, functioning as a reserve (base) node for another area (e.g., South America).

DNA BANK-NET NODES

WORKING NODES	RESERVE NODES
(Dispensing nodes)	(Base nodes)

Plant collecting, herbarium vouchers Initial field notes and ethnobotanical data	Plant collecting, herbarium vouchers Initial field notes and ethnobotanical data
Storage of DNA rich materials (leaves, shoot tips, etc.)	Storage of DNA rich materials (leaves, shoot tips, etc.)
Extraction and storage of genomic DNA	
PCR amplification of DNA using primers supplied by users Distribution of plant genes	

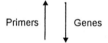

Primers   Genes

Biology, Biotechnology, Paleobotany users

FIGURE 13.2   Schematic representation of the flow of materials and the relationship between working (DNA-dispensing), reserve (base) nodes, and users.

General Requirements for Nodes in the DNA Bank-Net

The task group recommended the following minimum requirements for nodes (Adams and Adams 1991):

Working (DNA-dispensing) nodes

Personnel: Taxonomists/collectors, biochemists/molecular biologists, technicians for practical work, capable administration

Equipment: Storage facilities (liquid nitrogen, cryovats), extraction facilities (centrifuges, gel electrophoresis, UV spectrophotometer, etc.), DNA analyses and PCR duplication (PCR thermal cycler, microcentrifuges, etc.), distribution system (packaging and mailing supplies), computer (database for inventory and correspondence)

Reserve (base) nodes

Personnel: Technicians, capable administration

Equipment: Storage facilities (liquid nitrogen, cryovats), computer (database for inventory and correspondence)

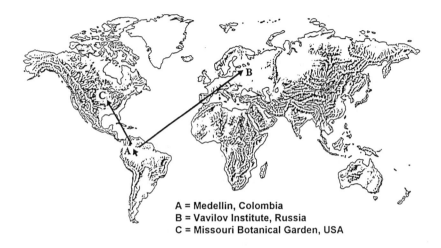

A = Medellin, Colombia
B = Vavilov Institute, Russia
C = Missouri Botanical Garden, USA

FIGURE 13.3   Hypothetical example of a triplicate collection in Colombia. The DNA-rich materials are placed in silica gel or Drierite for interim preservation and taken to the working node (A, Medellin). From Medellin (A), two replicates would be sent to reserve nodes at Vavilov Institute (B) and the Missouri Botanical Garden (C).

Each DNA collection should be split initially into at least two or three portions. One sample (DNA-rich material or extracted DNA) should be stored at a working (DNA-dispensing) node and another portion or portions be stored in at least one, but desirably two, back-up reserve (base) nodes. The reserve nodes should be in different countries and if possible on different continents to safeguard the DNA samples against various natural and man-made catastrophes. An example is shown in figure 13.3. where three replicate samples are collected and taken to Medellin with replicates and then sent to the Vavilov Institute and the Missouri Botanical Garden. Plant materials (in silica gel) could be stored in a freezer until the identification and other documentation have been accomplished and then shipped in quantity with other samples in off-season periods. No doubt other strategies will be developed with experience.

Several general recommendations that came from the task groups (Adams and Adams 1991) include:

a. DNA should be extracted from cryopreserved DNA-rich materials only when the DNA is needed. Delaying the extraction has the advantage of letting technology catch up so advanced techniques can be used as they become available.

b. Working nodes should generally be an existing organization with adequate biochemical expertise and have an associated herbarium. Having an herbarium on site would not be required but a very close, local (in the city) association with a recognized herbarium (Holmgren, Holmgren, and Barnett 1990) is required.

c. For the working as well as reserve nodes, it is necessary to have a strong institutional commitment, not just a personal commitment, in order that the collection be maintained in perpetuity, not just for the lifetime of one person who has committed him- or herself to the idea.

d. Consideration should be made concerning the availability of dependable electricity and liquid nitrogen in determining the feasibility of establishing a node.

e. Considerable interest was shown in the concept of storing composite DNA samples (e.g., a composite of DNA from all the legumes in a region, to be used for screening or retrieval of unusual genes).

f. The need for computer and database compatibility was expressed. Given the number of flat file and relational databases that are compatible with dBASE, it would seem that dBASE compatibility would be desirable. No consensus was reached with regard to this nor on the use of a flat file versus a relational database. It was felt that the critical issue at present was to begin collecting DNA-rich materials.

Scope of Plant Collections

The task group given this assignment felt there is a need for an initial focus rather than random collections and that economically useful plants should be given some priority (Adams and Adams 1991). However, this priority would not include the major crop plants of commercial usage that are widely cultivated (e.g., maize, rice, wheat, etc.) but rather those indigenous species that are tended or otherwise used by local people.

One problem with giving a priority to species is that field collecting then becomes "plant hunting" trips, which tend to be very expensive. It would seem that the cheapest and most practical way to preserve the largest percentage of plant genes would be to utilize the current (and additional) floristic collectors (such as those of the Missouri Botanical Garden, Royal Botanic Gardens, etc.), who are already in the field and familiar with the region's vegetation. The collections of DNA-rich material (leaves) could be done with little additional effort when specimens are collected.

Collecting Procedures

DNA collectors should be considered the same as all other plant collectors. Consequently they should (Adams and Adams 1991):

a. Voucher all collections in recognized herbaria (i.e., listed in Index Herbariorum, 8th ed [Holmgren, Holmgren, and Barnett 1990]).

b. Provide proper label information as to the locality, habitat, and so on, for each plant collected.

c. Follow all procedures concerning permits, convenios, and deposition of duplicate vouchers in the country of origin.

d. Collect leaf samples and pack them in desiccants (see Adams, Do, and Chu 1991) immediately (the same day). Leaves themselves provide simple, long-term storage.

e. In the case of legumes, samples of root nodules should be taken, if possible, but kept as a separate accession.

f. If a chemical treatment is used in the field, information should be provided concerning the method, and some untreated leaves must be stored in desiccant (see d above).

g. When possible, fossil material should be included in DNA Bank-Net. In this case, when destruction of the source material occurs, documentation via photographs and fragments is necessary.

h. Some material may be accessioned from herbarium specimens under the control of local curators using current methods of DNA extraction. Herbarium sheets should be marked if sampled for DNA. Herbarium specimens are limited in supply and appear to be useful only for material collected without chemical preservation. Material may be sampled directly from the sheet or the attached specimen envelope if it contains sufficient leaf material (ca. 0.1–0.5 g dry wt.) for DNA extraction.

Interim Field Storage of Specimens

The problems associated with bringing back fresh or frozen materials can generally be overcome by specialists (e.g., worldwide collections of fresh foliage of *Juniperus* for essential oil analyses and DNA by the author). However, botanists doing floristic research will likely collect many of the specimens from rare and endangered tropical species. They often collect specimens from scores of different species in a single day. The bulk of the materials they have to process and ship requires that any protocol for the collection of samples for specialized needs (e.g., DNA storage/analyses) must be quick, simple, and trouble-free. The generalist collector, working in tropical areas, cannot be expected to preserve hundreds of thousands of collections for months under tropical conditions and then arrange transport through customs, all the while keeping the individual specimens frozen.

Fortunately, at least as far as DNA preservation is concerned, interim preservation in silica gel or Drierite is an effective way to keep plant materials in the field or in transit for several months at ambient temperatures (Adams, Do, and Chu 1991).

Protocol for Field Preservation of Foliage

"Drierite" has a water capacity of 10 to 14%, but above 6.6% the capacity varies inversely with temperature (W. A. Hammond Drierite Co.). One would not want to risk possible rehydration of leaves, so storage ratios should be based on the 6.6% capacity. In lab tests, silica gel absorbed 8.85% of its weight of water after exposure to 100% humidity for 16 h at 22°C (Adams, Do, and Chu 1991). We have found that plant materials contain as much as 92% moisture (Adams, Do, and Chu 1991), so a useful approximation would be to assume that the plant is mostly water and use sixteen to twenty times the fresh leaf weight for the Drierite or silica gel component.

Now that inexpensive (U.S.$100) battery-powered, portable balances are available, one could take a supply of jars that hold, for example, 100 g of silica gel and then weigh out 5 g of fresh leaf material and add it to the jar along with silica gel (or Drierite). We have found that air-dried leaves (suitable for herbarium vouchers) generally contain from 10% to 15% water. Using a robust value of 20% water for air-dried leaves, one can weigh out 5 g of these leaves (5 g x 20% = 1 g water) or

1 g fresh leaves per 20 g of silica gel. This procedure may seem time-consuming, but in practice we merely do a quick check on the leaf area needed to give approximately 1 g (fresh leaves) or 5 g (dried leaves) and then just use that amount of leaf area. For example, for spinach, a fresh leaf area 2 x 4 cm weighs about 1 g. So one can just cut the leaves into roughly 2 x 4 cm squares and add one square to 20 g of silica gel. For succulent leaves a slightly different protocol may be used. Liston et al. (1990) removed succulent leaf material after twenty-four hours in Drierite and placed it in fresh Drierite. A note of caution is necessary concerning field drying of specimens for subsequent silica gel/Drierite storage. We have experienced difficulty obtaining DNA from leaves dried at temperatures higher than about 55°C. In very rainy conditions where high drying temperatures (from butane stoves, for example) are used to dry specimens, it would seem advisable to merely blot leaves free of surface moisture and then place the fresh leaf material directly into silica gel or Drierite. Liston et al. (1990) took 2–5 g of plant tissue and wrapped it in tissue paper to prevent it from fragmenting, then placed it in a 125-ml Nalgene bottle, 1/3 prefilled with Drierite (with blue indicator crystals), and then filled the bottle (2/3) with additional Drierite.

Clear plastic bottles are probably preferable to glass to avoid breakage in transit. Using clear jars allows one to check the indicating crystals without opening the jar. The lids should be sealed with vinyl tape to insure against moisture leakage. The use of parafilm to seal containers is not recommended, as we have found it to come loose at 37°C (and of course at tropical temperatures!).

Silica gel and Drierite do differ in one characteristic that may be a consideration. We have found that silica gel can be dried (recharged) at 100°C for 24 h, but Drierite must be dried at a much a higher temperature (200°C). In addition, we could easily dry (recharge) silica gel but were unable to dry (recharge) Drierite in a microwave oven. If the desiccant gets wet before use, silica gel appears to be much easier to dry. Silica gel is used in large quantities for flower drying and thus may be cheaper, depending on the source. Both Drierite and silica gel could be recharged for reuse on subsequent trips, but one should be very careful to remove any leaf fragments. If materials are to be checked through customs, it is useful to have a small container of silica gel/Drierite that you can open and show customs agents. A demonstration that the blue indicator crystals will turn pink when you breath on or moisten them is helpful in convincing the customs officials not to open your sealed specimen jars.

Future Considerations

The vast resources of dried herbarium specimens in the world's herbaria (see chapter 4) may hold considerable DNA that would be suitable for polymerase chain reaction (PCR). It seems likely that the integrity of DNA would decrease with the age of specimens. Because there are many types of herbarium storage environments (see chapter 15), preservation, and collections, there is a need for systematic investigations of the effect of modes of preparation, collection, and storage on the integrity of DNA in the world's major holdings.

One of the major concerns in storing DNA from extinct species is the limited amount of DNA available for distribution. A general process is needed by which the DNA could be immobilized, and then specific genes or oligonucleotides amplified. Genomic DNA onto nylon might be used as described by Kadokami and Lewis (1990) for cDNA from spiders. Amplification would then involve removing the membrane with the bound DNA from cryostorage and amplifying the desired gene, washing away the primers, and placing the bound DNA back into cryostorage. Although Kadokami and Lewis (1990) reported successful PCR amplification of membrane-bound cDNA, we have not been able to extend their work to genomic plant DNA. Additional research is needed in this area.

Research is also needed to amplify the entire genome DNA of a species. Some modification of the GAWTS type (Genomic Amplification with Transcript Sequencing; Sommer et al. 1990) protocol needs to be developed for eventual supplementation of DNA reserve stocks to obviate the need for replenishment from outside sources.

Technical workshops need to be conducted in order to bring experts together and develop specific techniques and protocols for DNA extraction, amplification, and storage.

Funding for initial start-up is modest for the reserve nodes but would require substantial funds for working nodes. Working nodes will need to be added to an existing molecular biology laboratory. Operating funds would be modest for a reserve node, depending on the cost of liquid nitrogen. Several international funding organizations may consider support for the establishment of nodes, but funding for operations will likely need to be borne by the institution or national government.

Acknowledgments

DNA Bank-Net has been supported by funds from the Conservation,

Food, and Health Foundation, the Helen Jones Foundation, and the Wallace Genetic Foundation.

Literature Cited

Adams, R. P. 1988. The preservation of genomic DNA: DNA Bank-Net. *Am. J. Bot.* 75:156 (Abst., suppl.).

____. 1990. The preservation of Chihuahuan plant genomes through in vitro biotechnology: DNA Bank-Net, a genetic insurance policy. In A. M. Powell, R. R. Hollander, J. C. Barlow, W. B. McGillivray, and D. J. Schmidly, eds., *Third Symposium on Resources of the Chihuahuan Desert Region*, pp. 1–9. Lubbock, Tex.: Printech Press.

Adams, R. P. and J. E. Adams. 1991. *Conservation of Plant Genes: DNA Banking and In Vitro Biotechnology*. New York: Academic Press.

Adams, R. P., N. Do, and G. L. Chu. 1991. Preservation of DNA in plant specimens by dessication. In Adams and Adams, *Conservation of Plant Genes*, pp. 135–152.

Barton, K. A., H. R. Whiteley, and N. Yang. 1987. *Bacillis thuringiensis*-endotoxin expressed in transgenic *Nicotiana tabacum* provides resistance to Lepidopteran insects. *Plant Physiol.* 85:1103–1109.

Booth, W. 1987. Combing the earth for cures to cancer, AIDS. *Science* 237:969–970.

Della-Cioppa, G., S. C. Bauer, M. L. Taylor, D. E. Rochester, B. K. Klein, D. M. Shah, R. T. Fraley, and G. M. Kishore. 1987. Targeting a herbicide resistant enzyme from *Escherichia coli* to chloroplasts of higher plants. *BioTechnology* 5:578–584.

Fischhoff, D. A., K. S. Bowdish, F. J. Perlak, P. G. Marrone, S. M. McCormick, J. G. Niedermeyer, D. A. Dean, K. Kusano-Kretzmer, E. J. Mayer, D. E. Rochester, S. G. Rogers, and R. T. Fraley. 1987. Insect tolerant transgenic tomato plants. *BioTechnology* 5:807–813.

Hawkes, J. G. 1979. Genetic poverty of the potato in Europe. In A. C. Zeven and A. M. van Harten, eds., *Proceeding of the Conference: Broadening the Genetic Base of Crops*. Wageningen, The Netherlands, 3–7 July 1978. PUDOC, Wageningen.

Holmgren, P. K., N. H. Holmgren, and L. C. Barnett. 1990. *Index herbariorum*. Part 1: *The Herbaria of the World*, 8th ed., *Regnum Veg.* 120.

Kadokami, Y. and R. V. Lewis. 1990. Membrane bound PCR. *Nucleic Acid Res.* 18:3082.

Lefebvre, D. D., B. L. Miki, and J-F. Laiberte. 1987. Mammalian metallothionein functions in plants. *BioTechnology* 5:1054–1056.

Liston, A., L. H. Rieseberg, R. P. Adams, N. Do, and G. L. Zhu. 1990. A method for collecting dried plant specimens for DNA and isozymes analyses, and the results of a field test in Xinjiang, China. *Ann. Mo. Bot. Gard.* 77:859–863.

Maiti, I. B., A. G. Hunt, and G. J. Wagner. 1988. Seed-transmissible expression of mammalian metallothionein in transgenic tobacco. *Biochem. and Biophys. Res. Comm.* 150:640–647.

Murai, N., D. W. Sutton, M. G. Murray, J. L. Slightom, D. J. Merlo, N. A. Reichert, C. Sengupta-Gopalan, S. A. Stock, R. J. Barker, J. D. Kemp, and T. C. Hall. 1983. Phaseolin gene from bean is expressed after transfer to sunflower via tumor-inducing plasmid vectors. *Science* 222:476–482.

Nelson, R. S., S. M. McCormick, X. Delannay, P. Dube, J. Layton, E. J. Anderson, M. Kaniewska, R. K. Proksch, R. B. Horsch, S. G. Rogers, R. T. Farley, and R. N. Beachy. 1988. Virus tolerance, plant growth, and field performance of transgenic tomato plants expressing coat protein from tobacco mosaic virus. *BioTechnology* 6:403–409.

Newmark, P. 1987. Trypsin inhibitor confers pest resistance. *BioTechnology* 5:426.

Plucknett, D. L., N. J. H. Smith, J. T. William, and N. Murthi Anishetty. 1987. *Gene Banks and The World's Food*. Princeton, N.J.: Princeton University Press.

Raven, P. H. 1988. Tropical floristics tomorrow. *Taxon* 37:549–560.

Redden, R. J., S. R. Singh, and M. J. Luckefahr. 1984. Breeding for cowpea resistance to Bruchids at IITA. *Protection Ecol.* 7:291–303.

Sommer, S. S., G. Sarkar, D. D. Koeberl, C. D. K. Bottema, J-M. Buerstedde, D. B. Schowalter, and J. D. Cassady. 1990. Direct sequencing with the aid of phage promoters. In M. A. Innis, D. H. Gelfand, J. J. Sninsky, and T. J. White, eds., *PCR Protocols*, pp. 197–205. San Diego, Calif.: Academic Press.

Shulman, S. 1986. Seeds of controversy. *BioScience* 36:647–651.

Ullstrup, A. J. 1972. The impacts of the southern corn leaf blight epidemics of 1970–1971. *Ann. Rev. Phytopathology* 10:37–50.

U.S. Congress. 1987. *Technologies to Maintain Biological Diversity*. Washington, D.C.: Office of Technology Assessment.

PART 5

Storage of Materials

Once plant materials have been sampled and preserved, the issue of storage becomes significant. A routine practice in the plant systematics community for centuries has been the exchanging of duplicate herbarium specimens, but in recent years this activity has come under increasing scrutiny by funding agencies. William Anderson in chapter 14 addresses these issues. For the first time we have a well-reasoned defense of this important practice, which we have taken for granted for so long. These perspectives are now documented for us and for our zoological colleagues as well as other interested parties.

Every herbarium has its own physical strengths and weaknesses. Some storage environments are clearly better than others, and some curators are more fastidious than others about watching for problems such as high humidity or pests. Christine Niezgoda in chapter 15 explores the question of how to grade overall storage environments for plant materials. At the very least, this should stimulate concern for improving our existing physical conditions and result in more focused planning of new facilities. We have little doubt that in the distant future herbarium materials will be accorded great historical value and be kept in constant temperature and humidity environments, such as currently exists in rooms that house rare books. But in the near future, how far do we want to go with these environmental storage recommendations, and is there a danger that someone outside the plant systematics community will seize on them in ways unpalatable to us? It seems obvious that we ourselves need to examine the issues closely.

14

The Importance of Duplicate
Specimens in Herbaria

WILLIAM R. ANDERSON

The collection of duplicate plant specimens is traditional among herbarium taxonomists. That practice is sometimes criticized, especially by nonbotanists, because of the real expenses associated with collecting and curating such duplicates. Taxonomists can justify some moderate number of duplicates, but they should do so on the basis of the scientific benefits conferred by them, such as documentation of variation in individuals and populations of plants. Duplicates also make collections available simultaneously to people with different needs working in different places. Taxonomists should also reconsider some of their traditional ways of interacting and their attitudes toward specimens, so that herbaria can maintain their independence in an era of increasing subsidies from external sources and the scrutiny that will come with such subsidies.

Plants lend themselves to the preparation of duplicate specimens, and from early times it has been traditional for plant collectors to take several instead of one whenever possible. In the eighteenth and nineteenth centuries, when lending specimens was not common practice and travel was difficult, expensive, and even dangerous, duplicates were the best way for taxonomists working in different locations to arrive at common concepts of taxa. It would be hard to exaggerate the importance of examining actual specimens oneself, now as then. Descriptions, databases, photographs, and scanned images are all helpful, but they pale in importance when compared to an opportunity to study the actual specimen that someone else has studied, and when that is not possible, studying a duplicate is the next best thing and still far superior to all the alternatives. Some of the best collectors have collected very large numbers of duplicates; for example, Cyrus Guernsey Pringle, whose excellent Mexican collections made him famous, collected in sets of sixty when possible (McVaugh 1972). I doubt that anyone is collecting that way today, but it is not uncommon for collectors to take sets of ten or more, and those duplicates continue to be accepted happily by herbaria around the world.

The reason herbaria continue to collect and exchange duplicates is that we rely on such exchange to increase the strength of our collections with speed and efficiency. If researchers at an institution want to build up a good representative collection of a particular group or from a particular area, they cannot possibly do it all themselves, but through carefully targeted exchange of duplicates much can be accomplished in a few years.

Before discussing the advantages and disadvantages of duplicate specimens, I need to be explicit about what I mean by the term *duplicates*. I am *not* referring to multiple collections of the same species, taken at different times and places so as to show variation in morphology and geographic distribution, and different stages in life-cycles and phenology. All biologists will agree that reasonable numbers of such multiple collections are essential. When I speak of duplicates here, I mean parts of a single gathering, made at a given time and place from a single individual or population.

The Problem with Duplicate Specimens

The problem with duplicate specimens is their cost. The cost of *any* specimen can be broken into three components. The first consists of collecting and preserving it, transporting it to the home institution, labeling it, and sending it to its ultimate destination. The second component consists of getting it into the collection of that receiving institution, which in the case of vascular plants comprises accessioning, mounting, and filing it. The third component consists of long-term curation. The specimen must be stored safely, retrieved when needed, sent out on loan, and restored to the collection on its return.

I do not intend to let myself be distracted here by attempting to assign firm numbers to these components; that would be exceedingly difficult and not essential to my argument. All that matters is to establish that the costs associated with all specimens (unicates and duplicates) are very substantial. One of my colleagues recently calculated that collecting a single specimen in Michigan costs between $1.00 and $5.00, and the cost in more distant places, especially foreign countries, is probably higher, although the number of variables involved confounds attempts at comparison. We calculate that mounting and inserting a specimen in the University of Michigan Herbarium costs approximately $5.00, and the whole transaction of lending a specimen, from request to return, must cost several dollars per specimen, especially when the loan is being mailed out of the country. Then there are the costs of constructing buildings, maintaining them year after year, and equipping them with such necessities as specimen cabinets, compactors, and computers. Some of these are one-time expenses that can be amortized over a certain number of years, but others, like climate control, are recurrent and perpetual.

Of course a substantial portion of the costs involved in curating and maintaining herbaria has to be assigned to causes other than the specimens. For example, the largest item in any herbarium's budget is salaries of researchers, and the cost of heat in winter is occasioned by the needs of frail humans, not dried plants. The same building that houses the plants houses the researchers, and most of the computers in a herbarium are used for research, not curation. The actual cost of curating a large herbarium is probably fairly low on a per-specimen basis. Nevertheless we cannot escape the fact that in a herbarium containing millions of specimens, the total annual cost of curating them is large.

Everything I have just said applies to all specimens, unicates as well as duplicates, so we need to focus now on the narrower question of the cost of duplicates. Locally this is not a major problem—modern herbaria, especially in this country, contain relatively few collections represented by more than one specimen, and our curators make a conscientious effort to avoid that kind of duplication. However, on a broader scale, the amount of duplication is extensive. Very old collections undoubtedly contain high percentages of unicates, but among collections made in the last 100 or 150 years, the percentage of specimens not represented by one or more duplicates in other herbaria must be fairly small, surely less than 50% and possibly less than 25%. Allowance must be made, of course, for the first specimen of any set; only n-1 of them are duplicates. Nevertheless, even after doing that, we are left with the fact that, whatever number we concoct for the real and fair cost of curating all the specimens in any herbarium, a major fraction of that number has to be considered the cost of curating duplicates.

Does it matter? Is this a *real* problem that scientists should worry about? That depends on who is paying the piper. If a herbarium has its own endowment or a generous higher administration, then what it accessions and curates is its own business, but few if any of us are that rich. Sooner or later, most of us are going to ask for public funds to support expenses that are at least partly curatorial, such as construction, equipment, supplies, and labor. We sometimes talk and act as if we were *entitled* to receive those funds without being subject to scrutiny and harassment by the watchdogs appointed to award them, but that is not a realistic expectation. People charged with spending public money cannot give it away without some basis. We cannot expect simply to account accurately for what we did with the government's largesse; we must be able to justify those expenditures.

Reasons for Collecting and Keeping Duplicates

First I feel compelled to point out an important difference between plant collections and animal collections, because our severest critics tend to be nonbotanists who do not appreciate some of the special problems associated with the study of herbarium specimens. Plants are morphologically extremely variable, both within and between individuals, and it is important to document that plasticity. In the case of small herbaceous plants, collecting duplicates offers a way to show the variation within a population, and a researcher who has access to such a set has far more information and potential for good biological insight than would be the case if he or she had only one or a few individuals. Duplicates of a single woody plant offer another scientific advantage. Developmental plasticity in plants often produces astonishing variation on different branches in features like leaf size and shape, inflorescence size and shape, number and maturity of flowers, and size, shape, and maturity of fruits. A series of duplicates documenting this kind of variation, which is what a specialist revising a group assembles through loans, enables us to produce much better science than would be possible with a single specimen constrained by the 29 x 42 cm of a standard herbarium sheet.

Aside from documenting variation, duplicate specimens confer many operational benefits. One obvious example is that redundancy ensures against disasters. Many of us have spent much time and effort seeking duplicates of specimens lost to the flames in Berlin, and in some cases where such duplicates are lacking we seem destined never to know just how an epithet should be applied or what a particular researcher's concept of a taxon was.

Redundancy can also resolve competing claims on available material. It is not rare for the needs of someone doing a revision to conflict with the needs of someone else preparing a flora. Duplicates make it possible for both projects to proceed simultaneously. Another advantage of redundancy is that it facilitates collaboration by people working at different institutions. For example, Scott Mori at New York and Iain Prance at Kew can work together on the systematics of Lecythidaceae because they both have duplicates of many of the same collections.

Monographers are the taxonomists who see the largest numbers of duplicates, and naturally those duplicates are sometimes more numerous than needed, but in addition to the advantages discussed above, a reasonable abundance of duplicates brings another blessing. One can borrow in the first instance from only a limited number of major collections and be confident of seeing representatives of most of the collections available. If collectors took only one or two specimens, far more borrowing would be necessary to approach the same completeness of sampling. Given the increasing cost of processing and mailing loans, the possibility of holding such requests to a minimum may become more and more important, especially now that the National Science Foundation is no longer willing to underwrite such operational expenses. Traveling to specimens is not a cheap alternative to loans or duplicates. It now costs so much to visit other centers, especially foreign ones, that such visits cannot possibly serve as a substitute for the slow and painstaking revisionary process possible at the home institution in the presence of a suite of borrowed duplicates.

Unfortunately, some of us, especially monographers, tend to forget that many people other than monographers use herbaria, and that their usage is often just as significant, both scientifically and quantitatively, as that of people preparing revisions. These uses are diverse, but they tend to share the characteristic that neither loans nor travel will substitute for ready access to a representative collection. Art Cronquist hardly could have written his books without having the treasures of the New York Botanical Garden at hand, and Rogers McVaugh and I never could have produced seven volumes of the Flora Novo-Galiciana in ten years without the marvelous collection at Michigan as our starting point. Indeed it is no accident that essentially all great floristic projects have been based at large herbaria that, with their rich mix of duplicates from many sources, make the floristician's task feasible. Then there are the many people who use herbarium specimens not as primary material for research, but as a reference tool—teachers, students, morphologists, ecologists, geographers, and others. Their needs are legitimate, but travel to a series of collections is almost never an option for them. Taxonomy is an end in itself for us, but we must not forget that we have an important service role to play, a role that often requires collections in which duplicates make the difference between success and failure in our ability to serve our colleagues.

Now I would like to offer some thoughts on collection of duplicates. As mentioned above, collecting is a fairly expensive activity, with much of the expense incurred by getting to the plants. Even if we do not rent helicopters, the cumulative cost of air fares, hotels, meals, assistants, and ground transportation is considerable. Compared to all those costs, the cost of collecting and transporting duplicates is negligible. After we finally get to a remote patch of forest in Peru or a quartzite outcrop in central Brazil, how should we behave? Obviously if we are constrained by time, transport, or other factors, we should collect only one or two sheets of each species. However, that is usually not the case, and usually it requires only a modest increment of time and effort to make five or six sheets instead of one or two, especially if we have cut a flowering branch that offers ample material for duplicates. So we are there, the material is good, and who knows when anyone may have such an opportunity again. The habitat may be gone in a year (see chapter 3 for an example of the destruction of tropical forests), and even if the plant survives, chances are good that it will not be in flower the next time a botanist gets there. Should we just walk away and leave that beautiful material to rot? Would that be a prudent use of collecting funds? There has to be a limit, but it is not obvious that the limit should be as low as one or two sheets.

How many duplicates should one collect? The minimum has to be three, especially when collecting in a foreign country—one for the host country, one for your own institution, and one for the specialist who is going to name it for you. I consider the optimum to be about six—the three mentioned above, plus three for deposition in other institutions. A set of six specimens usually allows one to convey an adequate sense of the variation and increases the burden of processing and transporting the specimens only moderately. I seldom collect larger sets now, but there are cases in which they are certainly justified, such as types of undescribed species or rare and especially interesting plants that will soon be destroyed by "development."

Recommendations

1. Herbarium botanists should not accept collecting protocols that would damage our scientific mission. If we can defend collecting and curating a certain number of duplicates, and I think we can, then we should continue that level of activity. However, we must be prepared to make that defense rationally, not on the basis of tradition but on the basis of the benefits bestowed by that level of duplication.

2. Institutions with active collecting programs should forge new or better coordination between collectors and curators. Collectors should not make more specimens than the home herbarium needs or wants; they are a cost as well as an asset, so collectors should be disabused of the notion that they have a right to collect all they want. For example, one tropical program not only collects excessive numbers of poor or mediocre specimens but regularly takes the same species five or more times in the same place at the same time. I name all the specimens of Malpighiaceae received, keep the few best, and discard the rest, but clearly the world's herbaria are being inundated with a series of specimens of dubious value, and the blame for this abuse has to be shared by both the collector and the curators at the home institution.

In a similar vein, collectors need to be well trained and closely supervised by someone who knows the flora being collected. The more the collector knows, the more likely he or she is to take a high percentage of plants that significantly contribute to systematic knowledge. A good example of such a collector is my Brazilian friend Gert Hatschbach, whose experience and amazing eye for novel plants enable him to collect the rare and interesting and leave the commonplace behind.

3. We should address the problem of voucher specimens. Herbaria cannot afford to curate vouchers for conservation, ecological, ethnobotanical, and other studies if the quality of those vouchers is so low that they have little or no value beyond their status as vouchers. Such specimens seldom come with any subsidies and certainly never bring enough of a dowry to underwrite their long-term curation, so herbarium curators need to seek some completely new approach to the whole matter. In principle this is not a part of the problem of duplicates, since vouchers are often unicates, but the two problems are inseparable, because they both revolve around curatorial costs. Most taxonomists would argue that judiciously selected duplicates are more valuable than many vouchers, so if we are going to cut back on acquisitions, it would be better to cut vouchers before duplicates.

4. Another way to make room for duplicates in herbaria and budgets would be to discard the least valuable specimens already present in herbaria. This is fraught with peril, of course, because a nonspecialist can seldom evaluate an apparently poor specimen in terms of its possible status as a type or in terms of its rarity or biogeographic significance. However, a monographer is uniquely equipped to do just that. When specimens go on loan to monographers, we should ask them to flag candidates for discard when returning the loan. If we discarded 5% of every loan on its return, that would add up to a lot of specimens over time. Again, it is much easier to make an argument for keeping a good duplicate than a scrappy, poorly documented unicate from a locality already represented for that taxon.

5. We must modify our ideas about what constitutes a "good" herbarium. There is a traditional mind-set among herbarium taxonomists that the best measure of a herbarium's activity and viability is its rate of growth. Historically this may have been valid, because there was usually a close correlation between collecting activity and research activity, but that has changed. Today, large-scale collecting is mostly centered in a few institutions, and it is often uncoupled from productive research. Herbaria should now be evaluated more on their loan activity, research activity, publications, and students—factors that more directly reflect their contribution to our understanding of plant diversity—than on numbers of specimens accessioned per year.

6. We should reevaluate the whole exchange system that is driving the growth of herbaria. I do not know whether we *can* do that, given our cultural bias toward sacred institutions like private property, debt, and credit, but we should think seriously about this tradition because it

causes problems. Duplicates should be placed in herbaria for legitimate scientific reasons, such as the presence of a specialist or to increase a preexisting strength in a group or from a particular area or because that herbarium stands a good chance of surviving in an era when many small herbaria will not. Science is not served by having valuable specimens go someplace only because of numerical debts. Another way to look at exchange and debt is the distribution of costs. If you send me a sheet on exchange, you have the cost of collection, labeling, and mailing, but I have the much greater eventual cost of mounting, inserting, and long-term curation, so who should receive the "credit" in this transaction? True, it all evens out if the exchange is even, but if it is not, the recipient actually bears the higher costs! Will we come to a day when herbaria that serve as repositories for large numbers of duplicates will charge a fee to take them in?

7. Ownership of specimens should be reconsidered. Economics is creating orphan herbaria and driving us toward consolidation and regional planning. "Owning" something in danger of being discarded is not at all the same as owning something everyone wants. If we are going to make the best possible accommodation to this new reality, so that future herbaria will be able to serve the needs of science and society, we must move beyond considerations of thine and mine and start talking about the most rational redistribution of specimens, equipment, personnel, and funds. Because of their abundance and cost, optimal disposition of duplicate specimens will be an integral part of such discussions; that is obvious to anyone who has ever worked with orphan collections. If we are not to lose control of those discussions, we need to start now to plan for a future that will be different but need not be disastrous.

Plant systematics is a diffuse profession, with the distribution of collections and expertise reflecting history rather than planning. Given that constraint, and our traditionally inadequate budgets, we have managed to achieve a great deal, and much of that achievement has been possible because we have been left alone to manage our own affairs. Nothing motivates people like a consciousness of both independence and responsibility. Fortunately or unfortunately, much of our splendid isolation may end in the near future. Strategically placed policymakers are finally getting the message that the future of the biosphere depends in no small degree on plant diversity, and from that understanding to an appreciation of the significance of herbaria is a small step. With more appreciation will come more support, and with those dollars will come more scrutiny by well-meaning but ignorant people who will want to tell us what we should, can, or must do. Pressure not to collect duplicate specimens is merely the tip of an iceberg of interference and uninformed micromanagement by outsiders, and we cannot afford to ignore the danger. We had better take the trouble to compound our own prescriptions or we may have to swallow someone else's nostrums.

Literature Cited

McVaugh, R. 1972. Botanical exploration in Nueva Galicia, Mexico, from 1790 to the present time. *Contr. Univ. Michigan Herb.* 2:205–357, 2 maps.

15

Guidelines for Plant-Storage Environments

Christine J. Niezgoda

This chapter focuses on examining current standards and trends for collections storage and maintenance and ultimately for development of a set of criteria on which herbaria can be evaluated. The collections "environment" for storage, preservation, and conservation of botanical specimens encompasses a wide range of issues. The most obvious are the physical aspects of storage (macroenvironment), including climate and humidity control, lighting standards, the utilization and organization of space, and pest control. Materials used in preparation, handling and storage (microenvironment) are just as important for the long-term preservation of collections. This includes the quality of the papers, adhesives, folders, boxes, and storage cabinets that are in direct contact with specimens. Also considered under collection's care are the associated issues of documentation; access and retrieval of collections; and the interaction of personnel with collections for maintenance, conservation, and research purposes. The table is proposed as a useful tool to evaluate and rate a collection based on the standards discussed in this chapter.

Natural science specimens are unique in that the vast majority are collected for research. As it becomes more difficult or even impossible to obtain natural history specimens, safeguarding present collections takes on ever increasing importance. The need to protect the integrity of the specimens for scientific research, which may involve destructive sampling, must be balanced against the need for preservation. Type and numbers of collections may vary from institution to institution but their needs for storage and conservation are similar. Whether we examine institutions where collections storage may be separate but within a public building or free-standing research facilities, the same issues must be examined: climate control, light, pest management, specimen preparation, storage facilities, specimen handling, and documentation. To preserve the collection one must utilize appropriate storage systems; identify the main causes that contribute to deterioration and develop methods to control them; and ensure that the usage of collections minimizes the effects of chemical, biological, and mechanical deterioration. Although the approaches and applications may vary, the ultimate goal should be the same: the establishment of a safe storage environment for the preservation of collections for present and future research.

Environment

A number of factors contribute to provide a nonreactive storage environment for collections. Thomson (1986) provides an excellent background and overview on the museum environment, including sections on light, humidity, and air pollution. Older buildings that are multifunctional must achieve a delicate balance between the dual purpose of their existence as a public exhibit/teaching area and collections storage area. Climate control, light, security, and pest management are much more difficult to control in these buildings compared to free-standing research and collections institutions.

Building Integrity

An overall inspection of the building is a necessary first step to pinpoint potential trouble spots. Brick buildings may have loose or missing mortar that allows seepage to occur as well as providing a point of entry for pests. Roofs and skylights should be adequately sealed against the elements. Interior overhead piping should be avoided in collections areas and libraries whenever possible; ruptures in these systems are especially damaging to specimens and books. Floors should be checked for cracks and holes. The removal of old pipes for radiator systems often leaves openings in floors; these should be sealed. Any operable windows present should be screened with a fine mesh. Doors that open into storage areas should remain closed to minimize the danger of pest infestations into the collections.

Climate Control

Controlled temperature and humidity is vital to all natural history collections. The heating/cooling system within collections storage areas should keep temperatures and humidity fairly constant. Wide fluctuations in temperature and accompanying humidity variation causing specimens to contract and expand repeatedly are a prime cause of accelerated deterioration. The control of humidity is more important overall than the control of temperature. If too much moisture is taken out of a system, organic products such as paper become less flexible and their fibers become easier to break. By contrast, in very humid conditions direct physical damage is less likely to occur, but these are the conditions that suit the growth of molds (Thomson 1986). The use of a recording thermohygrograph, which monitors both temperature and humidity, should be standard procedure in collections areas.

Lighting Conditions

Lighting systems in the overall environment of storage and work areas should be assessed as to their utility, illuminating power, and effect on specimens. A balance needs to be achieved where these areas overlap in order to incorporate appropriate illumination that is not harmful to specimens but that is adequate for working conditions. The ability to illuminate only parts of the herbarium or work areas by separate banks of lights or individual switches is both economical and practical. In this way specimens are only exposed to light for short periods of time when the lights are in use. Fluorescent lights are preferred in storage areas but may prove inadequate for work areas. The use of small task lamps placed at work stations economically solves this problem. Common sense dictates many of the other obvious precautions that should be taken: collections should not have long-term exposure to unfiltered UV light; collections must not be left out on counters for lengthy periods of time; windows should be outfitted with shades. Fluorescent lights with UV filters should be used in collections areas where specimens are housed on open shelving, and these areas should be illuminated only when in use.

Pest Control

Because of its importance to collections care, the subject of pest control in herbaria has been widely researched. The future trend for pest control is a move away from chemical usage in collections and toward pest control methods that do not have a negative impact on the scientific integrity of specimens. A wealth of information is available on museum pests and various methods of nonpesticide control that are currently in use or being extensively researched. Strang (1992) has reviewed available literature relating to thermal mortality of museum insect pests in relation to both high-temperature and freezing techniques. Florian (1990) has studied the effects of freezing on natural history specimens. Other research has focused on using extremely low O_2 conditions for pest control (Burke 1993). The Field Museum Conservation staff has experimented with flushing collections with CO_2 in a sealed atmosphere. Literature is also available on portable fumigation bubbles. This type of research takes on added importance as commonly used fumigants become unavailable—for example, the director of the EPA has mandated that U.S. companies phase out, by the year 2000, the production and importation of methyl bromide (*Fumigants and Pheromones*, no. 30 [Winter 1993]).

With the advent of more and more health concerns and less and less usage of pesticides and fumigants, the importance of instituting an integrated pest management (IPM) program cannot be overemphasized (Odegaard 1991). This program incorporates the "coordinated use of pest and environmental information with available pest-control methods to prevent unacceptable levels of pest damage by the most economical means and with the least possible hazard to people, property, and the environment. The goal of the IPM approach is to manage pests and the environment so as to balance costs, benefits, public health, and environmental quality" (EPA definition, 1990).

A successful IPM program incorporates several elements:

The isolation of incoming specimens   The shipping and receiving area should be physically separated from the permanent collection.

Treatment program for incoming specimens   Freezing is a widely used and accepted method of treatment. Upright floor models are suitable for low-volume herbaria. In smaller freezers it is advisable to remove the contents of larger boxes into smaller stacks enclosed in plastic. When dealing with large volumes of specimens a walk-in freezer is preferable so that incoming shipments can be isolated quickly. A temperature probe placed in the interior of packages is useful to monitor the actual low temperature reached within the materials. Recommended freezing temperatures are -20°C for at least forty-eight hours. Suggestions have been made to prolong the exposure of specimens for as long as one week or to go through two cycles of freezing. Before unpacking specimens, they should be allowed to adjust to room temperature before handling; twenty-four hours is recommended. Long-term storage of unmounted material in a cold room is advisable to further isolate potential sources of pest infestation.

Correct identification of pests   The services of entomologists are invaluable as are training courses on pest management and manuals on pest identifications.

Monitoring of collections   Traps should be used and visual checks made of groups susceptible to pest infestations. There are a number of pheromone traps for specific pests (*Lasioderma serricorne* Fabr.; cigarette beetles) available from vendors (e.g., Insects Limited, Indianapolis, Indiana).

Schedule for pest control within the entire collection   Freezing of all collections on a rotational basis is recommended. Large-scale fumigation should only be carried out by trained technicians; buildings must be sealed adequately to prevent exposure to workers and the environment (see earlier comment on methyl bromide).

Incorporation of a routine housekeeping program   This involves providing for regular dusting, sweeping, and vacuuming of storage areas; removal of live plant materials; strict enforcement of restricting food items in collections areas; and modification of storage systems if advisable and practicable (Williams and McLaren 1990).

Treatment of infested collections   A variety of methods are available (see references above). If the infestation is localized to a small area (e.g., one herbarium case), refreezing the case contents is a practical solution. A thorough cleaning and vacuuming of the case should be done before reinserting specimens.

Security

All personnel having access to collections must be issued adequate identifications via employee photo ID or a prominently displayed visitor pass. Doors that separate public areas from collections storage should be either specially keyed, accessed by key punch codes, or activated by magnetic codes. Especially valuable collections may be further restricted by a computerized key card access system: card readers for specific areas are issued only to staff; each entry to the area is monitored by computer; some areas (strong room) require a personalized code and the presence of two people for entry, and are further monitored by cameras. This higher degree of security is most commonly used within extremely valuable collections.

Health and Safety

Conditions in collections areas should not compromise the health and safety of staff and visitors. It must not be forgotten or ignored that older specimens in the collection may have been directly treated in the past by toxic chemicals (e.g., mercuric chloride). Many publications summarize the common pesticides and insecticides in past and present use (Story 1985; Hall 1988). Waddington and Fenn (1986) list a wide variety of sources and references for health and safety issues in natural history museums. Although many practices used in the past to control insect problems have been discontinued because of health concerns, such as the use of paradichlorobenzene (PDB) and naphthalene in herbarium cases (Linnie 1990), residues still remain on the specimens, folders, and cabinet gaskets. Therefore workers in the collections should be made aware of this and safeguarded. Gloves or masks or both may be advisable or mandatory by OSHA regulations. (*Cautionary note*: Anyone who chooses to wear a mask must first undergo a respiratory stress test and be properly fitted for an appropriate respiratory device by qualified personnel.)

Disaster Plans

The identification of potential disasters such as power failures, fires, floods, earthquakes, and other weather-related incidents that might affect the collections is foremost. One can minimize the exposure of collections to many of these disasters by locating collection storage areas away from overhead pipes or basement areas that may flood, utilizing appropriate storage units, and stabilizing open shelving with end bars. The presence of preventative systems is advised: fire doors, alarms, sprinklers (dry system to minimize the danger of water-filled pipes leaking), and back-up generators. In the event of a disaster the identification of resources, personnel, and equipment necessary for salvaging and storing damaged collections is essential. A chain of command, responsibility, and notification needs to be established in dealing with all situations.

Physical Storage

Collections areas need to be designed with two goals in mind: (1) safe storage of collections; and (2) easy access for the user. The physical layout and organization of the collection should maximize the available space and enhance accessibility and retrievability of collections.

Space Utilization

With the expansion of collections, the question of space utilization predominates in many institutions. Collections should always be placed within areas that are solely dedicated to storage; the use of public and unsecured space should be avoided. Older buildings often cannot be expanded because of lack of additional land or architectural constraints to preserve the integrity of landmark buildings. Installation of compactors where the additional weight is not a problem is often an economical solution that increases floor space without major renovation by eliminating fixed aisles. Compactor units can be designed to incorporate existing cabinets to cut down on replacement costs. Presence of adequate counter space, microscopes, electrical outlets for computers, efficient lighting, and step stools all contribute to ease of use.

Organization of Collections

Collections should be organized for ease of use by staff, researchers, and visitors as well as for the physical integrity of the specimens. Overcrowding of specimens in cases should be avoided as tight packing can cause compression, distortion, and breakage. Separation and segregation of different types of collections is both a practical and functional necessity in storage. Although the majority of mounted botanical specimens will be stored in standard herbarium cases, within genus folders, ancillary collections may have differing needs.

Ancillary collections   In these oversized collections, portions of specimens such as large fruits, roots, and bark usually cannot be successfully attached to herbarium sheets. If sheets are too bulky and are placed within genus folders, physical distortion of adjacent sheets occurs. Specimens can be incorporated into storage cabinets within the main collections area or segregated into specialized cabinets with drawers.

SEED COLLECTIONS. These collections, used solely for comparative purposes, are easily stored in clear glass vials with cork stoppers or small jars. Identifying labels on acid-free paper can be placed within the vials and are visible for filing and reference. Vials and jars can be stored in cabinets with drawers.

ECONOMIC BOTANY COLLECTIONS. These encompass a wide variety of forms and storage containers and are best organized as a unit, separate from the standard herbarium collection.

TOXIC/HALLUCINOGENIC MATERIALS. These need to be stored separately to ensure their safety and reduce the risk of exposure to workers. Appropriate logs should be kept to monitor the usage or consultation of these types of collections.

WET COLLECTIONS. These need to be kept in separate well-ventilated storage, away from the herbarium area. The collection should be actively monitored for liquid levels, leakage, and integrity of the container and seal. Appropriate fire precautions must be observed.

PHOTOGRAPHIC NEGATIVES, PRINTS, AND ILLUSTRATIONS. These are important parts of many collections that are often overlooked and stored improperly. Map cabinets are ideal for storing illustrations flat and secure from light and dust. Negatives need to be stored in a climate-controlled area and segregated into film types. Clamshell-type archival boxes are useful for print collections.

Cross-referencing ancillary material is imperative as this increases their use and ensures that they are not overlooked by researchers.

Other collections that are useful and practical to segregate from the remainder of the herbarium include:

PRESORT CABINETS. These house newly mounted materials sorted according to family. This allows for the organization of an efficient filing system since slots that are full can be spotted and retrieved for filing. Also, specimens within a family can be easily located for visiting researchers. Any materials that require special attention can also be placed in these cabinets (e.g., types to be verified, determinations to record, and specimens that need to be stamped or numbered).

CURATOR CABINETS. These house all newly mounted specimens and other incoming materials that are important to their research interests.

UNMOUNTED COLLECTIONS AND EXCHANGE MATERIAL. If possible, these specimens should be housed apart from the main collections or at least localized in one area (a cold room is ideal) for pest control. Since these collections are not checked as frequently as others, infestations may occur; isolation would control spread into the main herbarium.

INCOMING LOAN MATERIALS. Segregation of other institutions' materials ensure that their specimens are not inadvertently filed within the herbarium.

RETURNED LOAN MATERIALS. Separate housing of specimens is essential until they can be thoroughly checked for needed repairs, missing packets, and any annotations that are tracked by researchers.

Retrievability

The ability to locate and retrieve specimens in a timely fashion depends on the establishment of an efficient filing system. The following guidelines are commonly used in well-organized herbaria:

Organization by alphabetical sequence/phylogenetic sequence   Each system has advantages and disadvantages; ease of filing and retrieval is enhanced in an alphabetical system, whereas ease of identification of related taxa is enhanced in a phylogenetic system.

Color coordination   The use of colored folders for specific geographic areas enhance the filing and retrieving of specimens. Imprinting of the folders with the specific geographic area further minimizes filing errors (e.g., orange = South America).

Drop tags for materials on loan   These alert researchers to other workers in the field and allows filers to assess the space necessary for loan return.

Drop tags for unmounted materials   These record information as to country, year of accession, and institution and enable quick prioritization for the mounting and retrieval of specific specimens.

Drop tags for exchange and gift materials   Specimens that are to be sent out are identified by institution, researcher (gifts for identification), and priority for shipment. Shipments that are not urgent can be kept until a larger number accumulates or until a loan is sent to that institution.

Accessibility Guidelines

Access to collections should be strictly limited to staff and qualified researchers. Visitors should be given an orientation to the layout of the collections area in general. Written guidelines on the use of collections, description of the collections and their locations, useful information about the institution, loan policy, and so on, should be distributed to first-time visitors. Before collections are used, a visitor's knowledge of specimen handling should be ascertained. The visitor should be provided with work space, a cart for moving specimens between the herbarium and the work area, a microscope, annotation labels, and any other needed equipment. Policies on photocopying of specimens, destructive sampling (removing pollen or leaf samples), and refiling of specimens should be discussed. Two forms that are useful for visitors include:

Loan request form (English and Spanish)   A preliminary request for a loan that remains with specimens selected by the researcher until an official letter is received from the institution; ensures that the loan material segregated by the borrower is kept separate.

Visitor survey form (English and Spanish)   Tracks usage of collections by geographic areas, taxa, institutions, and personnel.

Materials: Preparation and Storage

Preparation of specimens is critical to their long-term preservation. Various mounting techniques are employed by herbaria for the typical specimen: sewing, strapping, gluing, and combinations of these three methods. Specialized techniques have been developed for unusual collections such as palms and cacti. The crucial component of all the methods is the use of safe archival supplies.

Papers

Archival herbarium paper (0.12 to 0.15 caliper) has a moderately textured surface for excellent adhesion properties and is buffered with 3% calcium carbonate to inhibit acid migration. It should be coded preferably with an institution's name or logo and sequentially numbered. Older specimens in the collection are often on poor quality, flimsy paper and whenever possible should be remounted to sturdier stock. Inclusion of an archival (acid-free and buffered) fragment packet is standard on all sheets to ensure that excess plant parts will not be lost or discarded. If packets need to be clipped for closure, plastic-coated (not metal) clips should be utilized.

Adhesives

The adhesives used should be compatible with archival standards and safety issues. Glues containing toluene (Archer's plastic adhesive), often used to "strap" plants, should be avoided both as a health hazard and because of their tendency to become brittle and crack. The polyvinyl acetate (PVA) glues have gained widespread acceptance because of their ease of application (sprayed or spread) and reversibility with water. (PVA is a fast-setting, acid-free adhesive that dries clear, can be remoistened, is water reversible, has excellent lay-flat properties, and does not become brittle with age.) Much research has been done on these types of adhesives. Baer, Indictor, and Phelan's (1975) evaluation of the PVA glues is especially pertinent as it considers the adhesive and the paper as a system. More recently, at the Canadian Conservation Institute, testing has focused on the effect of aging on the pH of PVA adhesives with acceptability ranges for cellulosic materials (Down et al. 1992). When modifying PVA glues by dilution it is important to monitor specimens for adhesive properties and to avoid excessive specimen "curl" upon drying. Weighting down the specimen and appropriate drying time are also critical to ensure proper preparation of the sample. Strapping should be done with a neutral pH gummed linen cloth tape. The adhesive on the tape, with a neutral pH, assures maximum adhesion as well as eliminating the potential of acid migrating to the mounting paper to cause yellowing. Precut strips are fairly expensive; rolls of tape cut by a paper cutter are much more economical and allow customized sizing of the various widths to accommodate strapping small delicate stems to large woody branches.

Storage Folders

Individual specimens are typically stored in plain or color-coded genus folders that are heavy-duty (.015 thick). The first choice should be acid-free and lignin-free with 3% calcium carbonate buffer to repel migrant acidity. Type specimens should be placed in individually labeled acid-free folders.

Specialized Storage Containers

Special consideration should be given to house larger materials unsuitable for normal mounting procedures. Palm collections are prime candidates for other types of storage. One method is as follows: the bulk material remains unmounted but is placed within a specialized four-flap palm folder made of durable acid and lignin-free stock. Cor-X (one piece "fluted" inert polyethylene) insert is used for additional support within the folder. For larger specimens, palm material is placed into archival boxes constructed from acid-free (pH 8.5) board with 3% calcium carbonate added to buffer migrant acidity. Metal edges, used to increase stacking strength, are attached without adhesives. Small pieces of the specimen are placed in resealable plastic bags (.004 thick). Original labels are glued to a numbered sheet within the folder or box. Polyethylene sleeves attached to the archival boxes hold current taxonomic information and leave room for annotation slips. The folders or boxes are color-coded with colored circles to the remainder of the herbarium.

Archival boxes lined with Ethafoam (nonreactive expanded polyethylene) inserts can be utilized successfully to preserve the integrity of Cactaceae collections. Appropriately sized resealable plastic bags are usually adequate for the majority of fruit collections.

Ethnobotanical material has a variety of special needs. Such collections can consist of jars, boxes, bags, and various-sized ethnographic materials. In an open storage system thin rolls of ethafoam can be used to line shelves. Ethafoam containers can be specially designed to hold glass jars. Acid-free boxes, tissues, and tubes should be used for fragile ethnographic materials. Special storage mounts can be constructed as needed for specialized collections. Rose and de Torres (1992) have compiled an excellent source of information on storage problems and solutions.

Storage Cabinetry

Any new cabinetry for collections storage should be constructed of nonreactive materials and have a proven resistance to fumigants and chemicals commonly used in a museum environment (i.e., material is chemically stable and does not off-gas or physically degrade to produce ureaformaldehyde, free sulfate radicals, sulfides, free sulfur, chlorides, acetates, chlorine, formaldehyde, oxides of nitrogen, oxides of sulfur, ammonia, organic acids, disodium phosphate, dibutyl phalate, acid-hardened phenol formaldehyde resins, peroxides, volatile organic compounds or plasticizers lacking long-term stability). Cases should have watertight tops to ensure the shedding of any moisture that might fall or collect on the surface from leaky pipes or sprinklers. Doors should be outfitted with durable hinges that open a full 180 degrees so as to lie flat against adjacent cases. A silicone gasket system provides protection for dust, light, and insects. Steel is degreased, phosphatized, and sealed before painting. The paint is nonreactive, has a solvent-free powder coating, is electrostatically applied, and then is baked. Cabinetry can be custom-made to fit special collections needs with repositionable shelving, drawers, and so on. Older metal cabinets should be checked for potential leaks from deteriorating gasket systems and replaced with new materials as needed. For economic botany collections the preferred storage method is in closed cabinets. The overall expense of such a system for a large collection often dictates the use of open shelving. Materials stored on an open-shelf system should incorporate a restraint system across shelving fronts to prevent accidental dislodging of materials.

Handling and Shipment of Specimens

Appropriate packing materials must be employed in all phases of specimen handling to ensure their safe transit to other institutions. Before sending specimens on loan, each sheet should be physically checked for needed repairs and the presence of archival packets. Specimens should be interleaved with newspapers or unprinted newsprint (the cost of using archival paper is prohibitive at larger institutions); bubble pack is advised on delicate or bulky parts as additional protection. Sheets are bundled in small stacks and completely wrapped and labeled; each wrapped package is further interleaved with cardboard to prevent damage. Types are placed in separate four-sided folders. The inclusion of a loan policy/guideline sheet outlining the institution's responsibility as to storage, handling, and annotation of specimens on loan is mandatory (Merritt 1992). It is important to use cardboard boxes of appropriate size with the addition of filler material to minimize shifting (e.g., polystyrene peanuts; since 1987 polystyrene peanut production has used no ozone-depleting CFCs; they are a clean and light-weight filler that is readily reused). Biodegradable peanuts made from starch that easily dissolve in water can also be used. However, long-term storage of these materials should be avoided as the starch may serve as a food attractant for pests. A copy of the invoice is placed within the box, and the original is mailed directly to the institution. Boxes are sealed and secured with paper strapping tape. The materials used should be recyclable or environmentally friendly in terms of its disposal.

Documentation

A thorough and detailed record documenting all transactions is essential throughout the entire process of specimen handling (Merritt 1992). Copies of pertinent and permanent documentation should be part of departmental records as well as the institutions archives. Types of documentation include the following:

Labels   Information relating to collection data from field notebooks should be kept with the specimen at all times until the label is permanently affixed to the herbarium sheet.

Barcodes   These enable tracking of individual specimens by a unique number on the herbarium sheet.

Accession records   These identify donor, type of transaction, collector, country of origin, and number of specimens.

Loan records   Documentation for the loan of specimens should be especially thorough and record the borrower (institution and researcher), the number of specimens, a listing of types, length of the loan, and any other special considerations.

Manual files   These are departmental copies of actual transactions, color-coded for domestic and foreign transactions, alphabetical by city, and color-coded for open and closed files.

Computer database   All the information in the manual file comprises the computer database which has the ability to quickly locate records for fields such as family, genus, institution, and researcher. It is useful for quickly generating a variety of data (monthly and annual reports, status of the loan, overdue loan letters, etc.).

Manuals   These include policies and procedures detailing each process; they are useful tools for training new employees.

Personnel and Resources

Well-trained and meticulous collections staff are an asset to any herbarium operation. The importance of staff to optimal functioning of a department cannot be overemphasized. All of us have had experiences with plants that have been inappropriately mounted—upside down, plant parts obscured, mismatched plants glued together. Specimens can easily be damaged because of careless filing or even lost because of misfiling. Aside from intense in-house training, opportunities for attendance and participation at workshops, seminars, or related courses should be encouraged and financially supported at all levels.

Information on conservation of natural history specimens is becoming more common and available because of the efforts of organizations such as the Society for the Preservation of Natural History Collections (SPNHC), the American Institute for Conservation (AIC), and the Canadian Conservation Institute (CCI). They also publish newsletters and journals that focus on issues of conservation and collections care. An excellent source of information on all aspects of herbarium practices and procedures is the recently revised *Herbarium Handbook* (Bridson and Forman 1992). Cato (1988) provides a summary of the extent of the professional structure that serves the needs of those who manage natural history collections. In addition, many vendors (e.g., University Products) recognize the niche to be filled by supplying archival materials. Great strides have been made over the past few years to increase the use of archival materials in products customized for natural history collections.

The potential contributions of other professionals, not specifically in natural history fields, should not be overlooked as sources of information. Object and paper conservators and librarians have expertise in working with paper materials and ethnographic objects, and they often can offer useful insights into applying conservation practices to biological specimens and suggest improved and innovative storage and design materials. Entomologists are invaluable for consultation and advice on pest problems and for critical identifications.

TABLE 15.1

Rating Herbarium Collections

This table provides a checklist for evaluating collections. Each element in the table is assigned a value from 0–2. Based on twenty-five elements detailed in the text, the maximum score is 50 points with the following ratings:

0–10 points Unacceptable
11–20 points Marginal
21–30 points Good
31–40 points Very Good
41–50 points Excellent

I. ENVIRONMENT—Collections Storage Area

Building Integrity	0—Major Renovation Needed 1—Minor Problems 2—No Apparent Problems
Climate Control	0—Major Fluctuations in Temperature 1—Moderate Fluctuations in Temperature 2—Steady state HVAC system
Lighting Conditions	1—Adequate Lighting 2—UV Filtered Lighting
Pest Control—IPM Program	0—None 1—Partial 2—Complete
Security—Passes and Doors	0—Uncontrolled Access 1—Partial Restricted Access 2—Strict Controlled Access
Health (Staff and Visitors)— **Protective Devices**	0—None Available 1—Limited Availability and Usage 2—Widespread Availability and Usage
Safety—Access to Controlled Specimens	0—No Controls in Place 1—Limited Access, No Monitoring 2—Strict Controlled Access and Monitoring
Disaster Plans	0—No Plan 1—Partial Plan 2—Complete Plan

II. PHYSICAL STORAGE

Space Utilization— **Location of Collections**	0—Space Inappropriate for Storage 1—Most Space Appropriate for Storage 2—Maximum Usage of Space and Location
Organization of Collections— **Ease of Use**	0—Haphazard Order 1—Limited to Mounted Specimens 2—Mounted and Unmounted Specimens
Retrievability—Specimen Tracking	0—No System 1—System for Some Specimens 2—System for All Specimens
Accessibility—Visitor Usage	0—No Guidelines 1—Oral Guidelines 2—Written Guidelines

III. MATERIALS UTILIZED IN PREPARATION, STORAGE AND PRESERVATION

Papers—Archival Standards	0—None 1—Minimal 2—Majority
Adhesives—Archival Standards	0—None 1—Minimal 2—Majority
Storage Folders—Archival Standards	0—None 1—Minimal 2—Majority
Storage Containers—Archival Standards	0—None 1—Minimal 2—Majority
Storage Cabinetry	0—Metal (old, leaking) 1—Metal (old, airtight/old, new) 2—Metal (new)

IV. HANDLING AND SHIPMENT

Materials—Appropriate and Recyclable	0—None 1—Partial Usage 2—Complete Usage
Methods—Standardized	0—None 1—Minimal 2—Majority

V. DOCUMENTATION

Methods—Specimen Tracking	0—Unorganized Manual Files 1—Organized Manual Files 2—Manual Files and Computer Database
Manuals—Detailed Procedures of Transactions	0—None 1—Some 2—All

VI. PERSONNEL AND RESOURCES

Collections Staff—Institutional/Grant	0—None 1—Understaffed 2—Appropriate Number
Conservation Staff—Consultation	0—Never 1—Rarely 2—Often
Auxiliary Staff—Consultation	0—Never 1—Rarely 2—Often
Information Resources—Journals, Organizations, Newsletters	0—Rarely Used 1—Sporadically Used 2—Widely Used

Evaluation

There is a need to develop a realistic set of standards on which collections handling, storage, and preservation can be evaluated. I offer such an evaluation rating system in table 15.1. By objectively analyzing and rating collections against a set of standards one can effectively target areas in need of improvement. Analysis of the needs and prioritization of those needs is necessary in implementing a plan for collections improvement. In addition, although a self-evaluation is a useful exercise, an external review that includes conservators, collections managers, and health and safety personnel is highly recommended. A critical assessment by an independent external panel can offer a fresh perspective, pinpoint problem areas, and propose potential solutions. It should be stressed that any criteria developed for judging and comparing collections should be used and viewed as a positive tool for improving a collection and not as a negative force.

Funding agencies are requiring that institutions seeking monies for collections support provide these kinds of information on collections assessments and institutional long-range plans for collections care. If one can provide a well-documented argument and a thorough plan for improvements in a collection, chances for funding should be more successful.

Literature Cited

Baer, N. S., N. Indictor, and W. H. Phelan. 1975. An evaluation of poly (vinyl acetate) adhesives for use in paper conservation. *Restaurator* 2:121–138.

Bridson, D. and L. Forman. 1992. *The Herbarium Handbook*. Kew: Royal Botanic Gardens.

Burke, J. 1993. Current research into the control of biodeterioration through the use of thermal or suffocant conditions. *American Institute for Conservation News* 18:1–4.

Cato, P. S. 1988. Review of organizations and resources that serve the needs of natural history collections. *Collection Forum* 4:51–64.

Down, J. L., M. A. MacDonald, J. Tetreault, and R. S. Williams. 1992. Adhesive testing at the Canadian Conservation Institute: An evaluation of selected poly(vinyl acetate) and acrylic adhesives. *Environment and Deterioration Report*, no. 1603. Ottawa: Canadian Conservation Institute.

Florian, M. L. 1990. Freezing for museum insect pest eradication. *Collection Forum* 6:1–7.

Hall, A. V. 1988. Pest control in herbaria. *Taxon* 37:885–907.

Linnie, M. J. 1990. Conservation: Pest control in museums—the use of chemicals and associated health problems. *Mus. Management and Curatorship* 9:419–433.

Merritt, E. 1992. Conditions of outgoing research loans. *Collection Forum* 8:78–82.

Odegaard, N. 1991. Pest control: Monitoring for insects in museums. *Conservation News* 11:14–17.

Rose, C. L. and A. R. de Torres. 1992. *Storage of Natural History Collections: Ideas and Practical Solutions.* Pittsburgh: Society for the Preservation of Natural History Collections.

Story, K. O. 1985. *Approaches to Pest Management in Museums.* Washington, D.C.: Conservation Analytical Laboratory, Smithsonian Institution.

Strang, T. J. 1992. A review of published temperatures for the control of pest insects in museums. *Collection Forum* 8:41–67.

Thomson, G. 1986. *The Museum Environment.* 2d ed. London: Butterworth.

Waddington, J. and J. Fenn. 1986. Health and safety in natural history museums: An annotated reading list. In J. Waddington and D. M. Rukin, eds., *Proceedings of the 1985 Workshop on Care and Maintenance of Natural History Collections,* pp. 117–121. Toronto: Royal Ontario Museum, *Life Sci. Misc. Publ.*

Williams, S. L. and S. B. McLaren. 1990. Modification of storage design to mitigate insect problems. *Collection Forum* 6:27–32.

PART 6

Conclusion

16

Are Present Sampling, Preservation, and Storage of Plant Materials Adequate for the Next Century and Beyond?

TOD F. STUESSY

At the present time 273 million specimens are housed in 2,639 herbaria throughout the world. With approximately 300,000 species of plants, this may seem a more than adequate sample. It is not evenly distributed, however, and an estimated 50,000 new species await discovery for which no material is yet available. The herbarium of the future (by the year 2050) will have all label data computerized. New data standards will be established for field collections, and these will be placed within geographic grids interfaced with Geographic Information Systems for maximum ecological correlations. Field expeditions will be highly collaborative with colleagues in host countries, with emphasis on long-term relationships. All specimens will be captured electronically as images, and botanical art will also be digitized. Botanical books will have all contents electronically available. These data will be on-line worldwide, thus greatly reducing the current information advantage enjoyed by workers in developed countries. Comparative data about plants will be formatted and available in SPE-

CIOSE, a textual and visual database of plant information. These data, plus environmental parameters, will form the basis for the BIO-SPHERE BOX (or BIOSBOX), which will permit access through computerized virtual reality. These innovations will allow herbaria to become more meaningful botanical resource centers for different patrons, including taxonomists, other biologists, and the general public. To prepare herbaria for the future, workshops should be convened now for community recommendations on (1) acquisition, maintenance, and ethics of use of DNA in collections; (2) standards for field data; (3) requirements for storage environments of collections; and (4) a national plan for regional storage centers for specialized plant parts (pollen, seeds, wood, DNA, etc.).

As anyone having read the preceding chapters of this book will attest, there are many dimensions to the sampling, preservation, and storage of plant diversity. So many in fact that it is a challenge to attempt a meaningful summary. This contribution is a commentary—a series of impressions that to me seems valuable and that I hope will be similarly viewed by at least some readers. I devote my attention to three issues:

1. How adequate is the plant sample in collections now, and what do we estimate it will be in the next century?
2. What will the herbarium of the future be like?
3. What specific steps might we take at this time in herbaria in anticipation of challenges we will likely face in the future?

How Adequate Is the Sample of Phytodiversity?

Before we can determine if the sample of plant materials in the world's herbaria is adequate, we must assess briefly what already exists. As Baum (chapter 4) points out, there are 273,000,000 specimens housed in 2,639 herbaria. In addition, there are other materials of wood, pollen, seeds, and isolated DNA retained in scattered collections (see chapters 10 through 13 by Carlquist, Roos et al., Loockerman and Jansen, and Adams, respectively), but these are relatively few in comparison with herbarium sheets. If we accept the existence of 300,000 known species of plants (Grant 1991), then this gives on average 1,000 specimens per species. In many cases this would be adequate to document simple patterns of variation in morphology and geography, but it probably would not cover aspects of ecology, pollination biology, intrapopulational patterns, hybridization, and so on. An obvious problem is that our existing sample is not average for all species; some taxa have been collected extensively, others scarcely at all. If we assume that 50,000 species of plants have yet to be described, then how can the sample be judged adequate if it is still so obviously incomplete (cf. similar comments by Heywood in chapter 1)?

I will dare to estimate that we will have 400 million herbarium sheets when the majority of sampling is over in the next hundred years. Access to plants in the remaining protected reserves will be limited—still allowing some access, of course, but stringently controlled. The sort of overprotection now practiced by Ecuadorean authorities in the Galapagos Islands, in which a permit is needed even to pick a few leaves from an abundant endemic species (such as *Scalesia pedunculata* Hook. f.; personal observation), will be the rule in all protected areas left, a minute fraction of the natural areas we now have.

Will this total sample be adequate? This depends on needs and purposes for which information from these materials will be extracted. Obviously there are real needs for botanical information from users outside our own discipline. Boom (chapter 2) outlines numerous constituents who need information from us, for which only specimens can help us answer. Other biologists, such as anatomists, biogeographers, ecologists, and evolutionary biologists, also need data about plants. Biographers, geographers, and historians are still other professionals who require data on plants. Many applied needs exist from engineers, developers, land managers, and politicians. Further, there is no lack of societal needs for data from botanical specimens (for a complete list, see Morin et al. 1989).

One of the problems of the botanical systematics community is our difficulty in focusing on needs. Heywood (chapter 1) emphasizes this point, indicating that we have never had a set of clearly defined goals. Development has been largely random and opportunistic. Witness the past two attempts to set priorities for systematic biology (Anonymous 1974; Stuessy and Thomson 1981), and read through these two "shopping lists" containing little prioritization. Predictably, few new funds resulted from these two previous substantial community efforts. The new *Systematics Agenda 2000* (Anonymous 1994) document *is* focused, setting high priority on inventorying the world's biota, but this may falter for lack of implementation.

Even without setting priorities we can address the need for information for botanical systematic purposes from the estimated final sample
of 400 million herbarium specimens. We will need data to help us
answer the following three questions: (1) What is the most predictive
classification for particular groups? (2) What is the phylogeny of these
groups? and (3) What have been the evolutionary processes that have
produced diversity in these groups? If we could assume that there will
be at least one informative specimen for all taxa, then we would be able
to make classifications and reconstruct phylogenies. But extinction,
especially fueled by human activities owing to increasing population,
may eliminate as much as 25% of the diversity (Raven 1990). It is almost
certain therefore that our final sample will not include some percentage
of recent plants. In geologically older groups this is a more serious problem because of natural extinction (e.g., Novacek and Wheeler 1992).
What percentage of missing data can we tolerate in the world's botanical basic data matrix so as not to seriously distort our increasingly quantitative assessments of relationships? An even greater problem, however, is to understand the biology of these taxa, for which one specimen
alone will likely be insufficient. We have a profound lack of understanding regarding character data for classification and reconstruction
of phylogeny. Sophistication of methods of analysis (e.g., NTSYS, PAUP,
Phylip, Hennig86, Component) has far outstripped the veracity of data
on which they are based. We need more ecological and life history studies of species so that their features can be understood in ontogenetic and
adaptive terms (e.g., Givnish 1986). We lack information on biotic interactions of structures with other organisms in all sorts of mutualistic and
competitive contexts. We have really only just begun to understand how
life forms interact in a detailed way. Hence the final sample of plant
materials will not be adequate to solve all these problems. We will still
have access to taxa in reserves, but this access will be strictly controlled.
Even then, the percentage of the world's existing plant resources contained in these remaining reserves will be a fraction of what we now
have, perhaps as low as 25 to 50%.

We must collect needed materials from the taxa that exist in remaining areas as opportunities present themselves (see an appropriate
example from the Philippine Islands; Sohmer, chapter 3). These will
include typical herbarium specimens as well as separate wood (dry or
pickled) and pollen samples (Carlquist, chapter 10) in prepared slides
or in liquid. Roos et al. (chapter 11) also show us ultra-cold methods of
pollen, seed, and bud preservation. Miller (chapter 5) reveals tech-

niques needed for collecting and handling bulkier specimens for pharmaceutical purposes. DNA samples are now being made specifically for genetic and evolutionary studies, and Adams (chapter 13) presents new developments with the DNA Bank-Net of institutions that are attempting to preserve these samples in an organized way. Loockerman and Jansen (chapter 12) show the potential of successfully extracting DNA from herbarium specimens. With PCR techniques and selected portions of the genone (such as ITS regions of rDNA), valuable phylogenetic data are present even in specimens more than fifty years old.

There is a clear message here: It is impossible that all institutions collect all plant materials—regional depositories must be established and a national plan developed. Obviously people who have already begun one type of storage option should be encouraged to continue it. More than one facility for each category of material is surely needed, with frequent sharing of materials to ensure that natural disasters or human miscalculations do not eliminate the only stocks of scarce material, which will continue to become even more valuable in the coming century. Anderson (chapter 14) stresses these numerous arguments with respect to exchanging duplicate herbarium specimens, which the botanical systematics community judged important many years ago. We remember the staggering loss of 4 million sheets in one evening on 1–2 March 1943 in the Berlin Botanical Garden and Museum owing to bombing and fire during World War II (Hiepko 1987) and realize the importance of spreading valuable plant resources to several regional centers. To house the many new specimens properly, we will need increased facilities with quality storage environments (Niezgoda, chapter 15).

The Herbarium of the Future

It might be profitable to cast a glance to the future to suggest what herbaria might be like in the year 2050 or 2100. Clearly if we knew where we were going we might be able to plan better in order to get there more efficiently, better prepared for inevitable changes and able to avoid certain pitfalls. It is safe to assume that more materials will be stored, and in different forms; but beyond that, what might we anticipate?

In view of the ideas presented in this book, my guess is that herbaria will be quite different a hundred years from now. To put this in perspective, let us review how herbaria have changed from 1894 to 1943 and then from 1944 to the present. This is an equivalent hundred years. In the first half of the past century (1894–1943), the plant systematics community concentrated on adding more specimens from more regions. The concept of voucher specimens developed, especially in relation to cytology and anatomy. Herbarium specimens were, in fact, used more routinely for anatomical information. In the latter half of this century more changes have occurred. Many more sheets have come from isolated, especially tropical, regions of the world. Pollen investigations have become a routine part of systematic work, and herbarium specimens have been viewed as a principal source of these data. Another most important development is the databasing of information from specimens already in collections (e.g., Brenan, Ross, and Williams 1975; Allkin and Bisby 1984; Bisby, Russell, and Pankhurst 1993).

As for physical facilities, in the second half of this century we have moved toward compactors. Perhaps we have been less effective in arguing for needed convenient space for our specimens (i.e., with aisles for efficient work), and have had to settle for compacted space. The view that compactors offer a better storage environment for specimens is a fiction; they simply allow more cases to be stored in smaller areas. If more space is realistically impossible, then compactors may be the best option—but they are surely not the ideal solution.

In terms of public outreach, we have begun to internalize the importance of interacting with the general public to justify our existence (Boom, chapter 2). In recent years some herbaria, particularly major ones such as the Missouri Botanical Garden and New York Botanical Garden, have become more skilled in spreading the message of the importance of botanical specimens for societal needs. The Botanical Research Institute of Texas (BRIT) is another positive example, perhaps even more so because it exists as a herbarium and botanical library in the absence of a botanical garden.

Despite these many changes, especially in the second half of this past century, I believe that the modifications in the next century will be even more substantial, perhaps revolutionary. Because herbaria are housed in different types of institutions, for example, universities, botanical gardens, and free-standing public museums, specific changes will not be the same in each. University collections place high priority on the generation of new knowledge through research, with a strong commitment to graduate teaching and lesser emphasis on public education and service. The larger the university collection, the more specimen loan activity (service) it will have. Local service in identification depends more on the herbarium's contacts with persons in the local area than on its size. A small collection may have many service obligations if such channels have been developed over the years. The botanical garden has a strong research focus, often horticultural, and is also committed to the informal education of children and adults; service also looms large. Free-standing public museums fulfill research, nonformal education (both for adults and children), and service roles (through identifications and loans to the rest of the systematics community). Nonetheless, despite these institutions' varying orientations, some developments will occur that will be common to all of them. These are the changes I wish to explore.

The herbarium of the future will have all data in a computer; data not yet in the computer will be viewed as data that is not worthwhile (essentially nondata). Data from newly acquired specimens will be placed immediately in databases as they are accessioned. This will include all label data (already done at Missouri Botanical Garden) plus comparative information generated later in the laboratory or garden plot. Observations made in the field will also be data-banked through handheld computers. As these data are recorded, they will be sent automatically via satellite to the home institution, to be edited upon the return of the field-worker (see Morain 1993, for suggestions in this direction). These data will be formatted and entered into the database at the time the labels are processed. Standardized community formats for these data will be used; data will meet minimal standards globally, or they will be judged of little value. This will help correct the problem of nonuniformity of data emphasized by Heywood (chapter 1). We are at present actually not too far from this capability (e.g., new geographic data standards; Hollis and Brummitt 1992).

Specimens will be in a geographic grid system worldwide. Old specimens will obviously not all be retrievable at the same level of accuracy, such as those with handwritten labels from the late 1700s with nothing more than "New Spain," but all new materials will be precisely located by latitude and longitude by Geographic Positioning System (GPS) devices, such as those commercially available even now. Geographical Information Systems (GIS) will provide numerous ecological correlations, and they will emphasize point data rather than polygons (DeMers, chapter 9). This will mean that decisions on obtaining new data will be facilitated by first determining electronically how many specimens are available worldwide for a particular taxon, how they are spatially distributed, and how comprehensive the existing sample appears to be for the specific project being contemplated. The impact on land-management decisions will be enormous.

All specimens will be captured as images and archived in some new high-capacity type of CD-ROM. Advanced scanning machines will be able to digitize images rapidly. These will be internationally available on-line through future generations of Internets. We will begin with type specimens or other special collections, as shown by the Ohio Flora visual archive project described by Kramer (chapter 7) or the innovative SMASCH project of California herbaria (Duncan 1991). This also means that workers in all countries will potentially have visual access to all botanical materials, hence substantially reducing the advantage that investigators in developed countries currently enjoy over those in other regions.

Future collecting expeditions in host countries will be totally cooperative and more in the context of research and educational projects of a long-term nature and of obvious benefit to host colleagues. We are thankful that the era of quick collecting "snatch-and-grab" expeditions to other countries is now gone; the importance of mutually beneficial interactions and development of scientific infrastructure is obvious (Stuessy 1991), and this will increase as the decades roll on.

Botanical art held by herbaria will be on CD-ROM (or video disc) and will be completely computer accessible worldwide. Note the ambitious current project described by White (chapter 6) at the Hunt Institute for Botanical Documentation to record electronically all original nonoil botanical art in public institutions and private hands (excluding prints). Inventory of illustrations for all taxa will be completed, and on-line images of all art will be available from herbaria.

Botanical books in our herbarium libraries (at least by the end of the next century) will be completely electronically captured and computer retrievable. By this I mean all contents, not just titles. If this seems unbelievable, the U.S. Library of Congress has just announced a plan to digitize 5,000,000 maps, pamphlets, and speeches and have these on-line by the year 2000 (DeLoughry 1994). A "virtual library" is in the offing (Cage 1994) and will be available worldwide. But this access will further lessen the intellectual gap between countries and result in much greater competition.

As a nomenclatural aside, when the botanical literature is completely electronically captured, all plant names will be conveniently indexed from these volumes, which will obviate the need for Names in Current Use, now championed so strongly by European botanists (e.g.,

Hawksworth 1991, 1992, 1993). The question is whether it is worth waiting rather than arriving at an immediate, but less than completely adequate solution. I favor working to gather all past literature for a more lasting and effective permanent nomenclatural resolution.

All other material stored in herbaria will be cataloged and listed or imaged in on-line form. This will include pollen samples, wood, pickled collections, dried material for DNA, and frozen samples. The quality of these materials for particular purposes will also be indicated, especially for the more sensitive analyses, such as with DNA and terpenoids. It is possible that a hundred years from now all DNA will have been removed from existing specimens and stored in a more stable fashion; this seems particularly likely following recent developments in DNA storage (Adams, chapter 13).

Collection environments will be closely monitored for critical dimensions of temperature and humidity, pest control, and so on, following guidelines such as those presented by Niezgoda (chapter 15). As specimens become less available in the wild, those in collections will become more valuable, and the quality of storage conditions will naturally improve correspondingly (one does not store the Hope Diamond in a shoe box in the closet!). White (chapter 6) recommends 68 to 70°F and 50 to 55% humidity as good conditions for preserving artwork, and presumably this would be suitable for literature and herbarium sheets as well. The important point is to avoid wide environmental fluctuations.

We shall all become better communicators with the general public. No collection will be immune from justifying its existence to those who pay the bills. As increasing levels of the human population force higher costs on property and space (witness the present situation in Japan), the need to justify each square foot of collection space to taxpayers of cities, regions, or country will be paramount. Herbaria will have more displays, more interactive computerized identification and educational modules, more workshop rooms, more auditoria, and more reading rooms for the general public. New herbaria will be designed more along the lines of local town libraries as resources for interested persons to enjoy, take advantage of, and become involved with. Herbaria in universities that plan for increased communication will reap commensurate benefits. Those that restrict their activities only to research and graduate teaching will find it difficult to survive a hundred years from now, unless they are essentially converted to federal research collections.

As access to new botanical materials in the wild becomes more restrictive, and with the capturing of images on computer, the need for duplicate specimens will diminish. Further, and more important, the specimens themselves will primarily have archival significance, consulted less frequently for normal research purposes (Morin and Gomon 1993). Video images will be useful for answering many of the questions now directed at original materials. In addition to these changes, I believe that two other advancements will have a major impact on our community (and the world). The first is the establishment of a database and visual archive of the world's flora, which we may call SPECIOSE. This may well be an extension of the present efforts to produce a new databased *Species Plantarum* for the world (e.g., Prance 1992). This will include all specimens, comparative data (chromosome numbers, anatomical data, pollen information, etc.), illustrations (both published and unpublished), geography, ecology, GIS plotting, and so on. In short, this electronic file will have all the available data for all the known plant species on earth. It will be computer driven and respond to our needs directly through specific queries we make. It will be the principal mode for the initiation of new research projects. All monographic work will begin with interaction with SPECIOSE for several weeks or months of examination and study. This will lead to a better circumscription of the problem, refinement of critical questions to be asked, sketches of preliminary relationships, data on field locations and ecology, and so on. In fact it will be possible to carry out many facets of revisionary work (e.g., Leenhouts 1968; Maxted 1992) without even leaving the confines of a single herbarium and the SPECIOSE database. Clearly this will not answer all questions, but it will allow many to be answered much more rapidly than before. This will also have tremendous value for informal and formal teaching. Data will easily be presentable from SPECIOSE on screen in lecture halls and auditoria (possible even now with LCDs on overhead projectors), which will facilitate communication about the botanical world. Nontechnical interfaces will be developed to enable the general public to access information from SPECIOSE.

But the developments will not stop there. Stredney (chapter 8) has described incredible current innovations in virtual reality that in the future will no doubt be truly astounding. Here we can envision the second major advancement, called the BIOSPHERE BOX or BIOSBOX for short. This will be a small room, accompanied by proper headsets for the user, which will give us entry to the biotic world stored electronically. Here we will walk through biomes of the world (e.g., savannas, deserts, scrubland, forests), dive under the sea, stand at the base of the tallest redwood, and be impacted simultaneously by visual, auditory, and olfactory senses of all data available about these organisms and ecosystems. We will be able to cut through a log with a virtual chain saw and count the annual rings of the tree or check the potential for lumber (assuming of course that these data have already been put in the system). We will be able to open a flower bud and examine the floral parts, all in virtual space. This will have powerful implications for conservation, education, land planning, development, and numerous other applications. Entering the BIOSBOX will be required for answering most questions about the planet. It will also reveal gaps in our knowledge. If we cut open a log with the virtual machete and find it blank inside, it will signal that we lack information on the anatomy of that species. Clearly the data of SPECIOSE will be the basis for that presented in a virtual way in the BIOSBOX.

All kinds of additional cultural applications can be envisioned with the BIOSBOX, such as in psychological therapy, classroom teaching, training programs, and so on. Young lawyers will get virtual trial experience in famous cases, testing their own ability to argue the facts. Criminals can be put in virtual situations similar to those in which they encountered their victims, to aid in their rehabilitation. The list is endless. The problem for society will be to establish limits on the use of this technology so that abuses do not occur. Virtual experiences could become so mind-distorting that persons might become addicted psychologically to the virtual world and lose their ability to cope with the real one, perhaps serving up worse societal problems than present substance abuses. The seductive aspect of the virtual world is that the user designs it to be what he or she wishes—it is driven by the mind of the user. The real world is not that way at all. Stredney (chapter 8) has said (paraphrased from Ganapathy), "Reality is a myth; all that matters is perception." This may be true, but it also harbors profound dangers if not managed correctly. We can never allow virtual reality to become the true reality or we will have lost control of our own destiny.

Specific Steps We Might Take Now

Different opinions will prevail on what steps might be taken in response to perspectives outlined above. Some people would favor more sophisticated databasing; others might stress more attention to storage environments. I have my own biases, as outlined below.

We need to deal most urgently with issues relating to DNA stored in herbaria. We need guidelines on how to treat old specimens and those still to be acquired. If we are to develop a "genetic insurance policy" (Adams, chapter 12), we need to deal with that part of the resource already available. Because herbarium specimens have been treated in so many different ways, the amount of quality DNA in different specimens will also vary. Some have been heated, others treated with alcohol or formaldehyde, some poisoned with mercuric compounds, and so on. Despite all these different techniques, usually no mention is made on the specimens. For extraction of DNA, it becomes largely a random process. We need to develop a label for each sheet indicating the treatment the specimen has received. Has it been heated, and if so, to what temperature and for how long? What chemicals, if any, have been used in its preparation? Proprietary rights of DNA in these specimens is yet another concern. This has rarely been problematic with other types of data, but with gene sequences, which might have future commercial value, who should be allowed to access the DNA, and under what circumstances? A workshop should be convened to address these issues and make recommendations to the collections community.

We should also convene a workshop to consider community standards for field data (and specimen data in general). What exactly do we believe will be needed now and in the future? This is obviously the time to make final decisions to guide remaining data-gathering at the end of this century and into the next. We may wish to establish the "Maximally Informative Field Record" (MIFR), but we need to know exactly what data should be included. Obviously, at the minimum, we will need precise GPS coordinate data, as well as habitat, ecology, and other types of biological and abiotic information. We should encourage pilot projects to learn levels of geographic precision that exist within our present collections, that is, how detailed are the locality data that now exist?

It would be valuable to have a workshop on the grading of storage environments, a topic opened for discussion by Niezgoda (chapter 15). What should the community do with regard to setting its own standards for collections maintenance? For the long-term good of the materials, it makes sense for us to begin this discussion now; it is also likely that external forces will eventually require us to do it anyway. The alarm of potentially having to capitalize museum specimen inventory, which came from the Institute of Museum Services but mercifully disappeared, is an example of what could happen if we do not address the storage issues ourselves.

Finally, we need a national plan for the preservation and storage of different plant parts, with emphasis on regional centers. We should build on existing specialized materials and create better collections for special needs, with proper curation and support. This should include the cold storage of seeds, pollen, buds, and even DNA. Who should be involved, and how will the needed costs be met? An overriding theme of this book is that we now begin to anticipate future roles of herbaria and to address issues pertinent to those roles. The sooner we do so, the more easily we will adapt to these new challenges, and the more effectively we will fulfill our obligations to society. It would behoove us all to heed an appropriate perspective offered by Heywood (chapter 1): "It is a privilege to hold plant specimens—not a right." Let us make the most of this privilege so that herbaria retain and increase their importance as guardians of plant diversity on this planet.

Literature Cited

Allkin, R. and F. A. Bisby, eds. 1984. *Databases in Systematics*. London: Academic Press.

Anonymous. 1974. Trends, priorities, and needs in systematic and evolutionary biology. *Syst. Zool.* 23:416–439.

Anonymous. 1994. *Systematics Agenda 2000: Charting the Biosphere. Technical Report.* New York: Systematics Agenda 2000.

Bisby, F. A., G. F. Russell, and R. J. Pankhurst, eds. 1993. *Designs for a Global Plant Species Information System*. Oxford: Clarendon.

Brenan, J. P. M., R. Ross, and J. T. Williams, eds. 1975. *Computers in Botanical Collections*. London: Plenum.

Cage, M. C. 1994. The virtual library. *Chron. Higher Educ.* 41 (4): A23, A27.

DeLoughry, T. J. 1994. Library of Congress announces plans to digitize several of its collections. *Chron. Higher Educ.* 41 (8): A41.

Duncan, T. 1991. *Development of a Specimen Management System for California Herbaria (SMASCH)*. Working Paper No. 1. Berkeley, Calif.: Assoc. Calif. Herbaria.

Givnish, T. J., ed. 1986. *On the Economy of Plant Form and Function*. Cambridge: Cambridge University Press.

Grant, V. 1991. *The Evolutionary Process: A Critical Study of Evolutionary Theory.* 2d ed. New York: Columbia University Press.

Hawksworth, D. L., ed. 1991. *Improving the Stability of Names: Needs and Options.* Regnum Veg. No. 123. Königstein: Koeltz Sci. Books.

Hawksworth, D. L. 1992. The need for a more effective biological nomenclature for the 21st century. *Bot. J. Linn. Soc.* 109:543–567.

_____. 1993. Protection des noms d'usage courant: un avantage ou une menace pour le taxonomiste. *Cryptogamie, Mycol.* 14:149–157.

Hiepko, P. 1987. The collections of the Botanical Museum Berlin-Dahlem (B) and their history. In H. Scholz, ed., *Botany in Berlin*, pp. 219–252. Berlin: Bot. Garten und Bot. Mus. Berlin-Dahlem. [*Englera* vol. 7.]

Hollis, S. and R. K. Brummitt. 1992. *World Geographic Scheme for Recording Plant Distributions*. TDWG, Pl. Tax. Database Standards No. 2, vers. 1.0. Pittsburgh: Hunt Institute for Bot. Documentation.

Leenhouts, P. W. 1968. *A Guide to the Practice of Herbarium Taxonomy*. Regnum Veg. No. 58. Utrecht: Intern. Assoc. Pl. Tax.

Maxted, N. 1992. Towards defining a taxonomic revision methodology. *Taxon* 41:653–660.

Morain, S. A. 1993. Emerging technology for biological data collection and analysis. *Ann. Missouri Bot. Gard.* 80:309–316.

Morin, N. R. and J. Gomon. 1993. Data banking and the role of natural history collections. *Ann. Missouri Bot. Gard.* 80:317–322.

Morin, N. R., R. D. Whetstone, D. Wilken, and K. L. Tomlinson, eds. 1989. *Floristics for the 21st Century*. St. Louis: Missouri Bot. Garden.

Novacek, M. J. and Q. D. Wheeler, eds. 1992. *Extinction and Phylogeny*. New York: Columbia University Press.

Raven, P. H. 1990. The politics of preserving biodiversity. *Bioscience* 40:769–774.

Stuessy, T. F. 1991. International collaboration in natural history museums: Organisms as international ambassadors. *Korean J. Pl. Tax.* 21:217–227.

Stuessy, T. F. and K. S. Thomson, eds. 1981. *Trends, Priorities and Needs in Systematic Biology*. Lawrence: Assoc. Syst. Collections.

INDEX